SPACE
In Science, Art and Society

Space is multiple. There are as many different kinds of space, and human engagements with space, as there are scales, media and cultures. In this volume, based on the 2001 Darwin College Lectures, eight distinguished authors from across the arts and sciences offer a panoramic view, inviting the reader on a fascinating journey of exploration. We begin with the human brain and the inner space it generates. We consider how space is used in sign language and in architecture. We learn of the virtual space created by computers and its artistic possibilities. We see how the demarcation of geographic space has profoundly shaped our past and present. Finally we contemplate outer space, its exploration and its ultimate nature.

SPACE

In Science, Art and Society

Edited by *François Penz*, *Gregory Radick* and *Robert Howell*

CAMBRIDGE
UNIVERSITY PRESS

PUBLISHED BY THE PRESS SYNDICATE OF THE UNIVERSITY OF CAMBRIDGE
The Pitt Building, Trumpington Street, Cambridge, United Kingdom

CAMBRIDGE UNIVERSITY PRESS
The Edinburgh Building, Cambridge, CB2 2RU, UK
40 West 20th Street, New York, NY 10011–4211, USA
477 Williamstown Road, Port Melbourne, VIC 3207, Australia
Ruiz de Alarcón 13, 28014 Madrid, Spain
Dock House, The Waterfront, Cape Town 8001, South Africa

http://www.cambridge.org

First published 2004

Printed in the United Kingdom at the University Press, Cambridge

Typeface Iridium 10/14 pt.   *System* LaTeX 2_ε [TB]

A catalogue record for this book is available from the British Library

Library of Congress Cataloguing in Publication data
Space / edited by François Penz, Gregory Radick and Robert Howell.
      p.   cm. – (Darwin college lectures)
Includes bibliographical references and index.
ISBN 0 521 82376 5
1. Space and time.   2. Space and time – Psychological aspects.   3. Space (Architecture)
4. Personal space.   I. Penz, François.   II. Radick, Gregory.   III. Howell, Robert.
IV. Series.
BD620.S68   2004
114 – dc22   2003055909

ISBN 0 521 82376 5 hardback

Contents

The plate section is between pages 88 and 89.

Introduction

FRANÇOIS PENZ, GREGORY RADICK AND ROBERT HOWELL

What is space? No single definition will serve. Space, after all, is multiple. Arguably, there are as many different kinds of space, and human engagements with space, as there are scales, media and cultures. A book about space could, therefore, be about almost anything, indeed almost everything. This book takes a deliberately inclusive approach. The chapters within were first delivered as lectures in an annual series at Darwin College, Cambridge University. The series, like the college itself, is interdisciplinary. Each year eight experts from across the arts and sciences are invited to address a given theme, before a general audience, from the distinctive perspectives of their fields. The lectures, and the now sizable shelf of books that commemorate them, aim to give large subjects such as colour, intelligence and evolution the panoramic treatment they deserve. In 2001, it was space's turn.

Why space in 2001? As organisers for that year, we found the match irresistible. For one thing, space is, in several senses, a natural counterpart to time, the subject of the millennial, 2000 series. For another, we, along with countless others, associated 2001 above all with Stanley Kubrick's seminal *2001: A Space Odyssey* (1968), voted in one recent poll the most influential and admired film ever. In the spirit of the film, we selected particular topics and speakers with an eye to transporting the listener – and now the reader – on a journey from inner to outer space, from consciousness to the Cosmos. In between, the ports of call include space and language, architecture, virtual reality, maps of the Earth, international politics and interplanetary exploration. Here, then, is a space odyssey for the intellectual traveller of broad tastes, who likes to be informed, challenged and entertained.

Space: In Science, Art and Society, edited by F. Penz, G. Radick and R. Howell.
Published by Cambridge University Press. © Darwin College 2004.

The journey begins with the most personal and in many ways puzzling spaces we know: our own minds. In her chapter on 'Inner space', the neuroscientist Susan Greenfield asks what science has learned about consciousness and its basis in the brain. Consciousness, she points out, is not an either/or property, but rather something adult humans enjoy in greater or lesser degrees, depending on circumstances. There is all the difference in the world between the intense focus and self-awareness of a mind sitting an exam, say, and the near obliteration of these qualities in a mind on Ecstasy. Greenfield argues that underlying this continuum of consciousness is a continuum of states of brain connectivity. The stronger and more numerous the connections between different parts of the brain, the greater the degree of consciousness. Of course, as she insists, the discovery of this apparently simple correlation between brain states and mental states leaves untouched the problem of understanding *how* brain activity causes mental states. But even without such understanding, the neuroscience of consciousness may be edging closer to a new generation of therapies for some of the most devastating mental illnesses, including Alzheimer's.

Between the private space of the individual mind and the public spaces of the wider culture is language. The relations between mind, language and culture have long been contested. Do the thousands of different human languages provide roughly equivalent windows on the world? Or, on the contrary, does each language determine a unique, culturally specific 'thoughtworld' (as the linguist Benjamin Whorf put it)? The psychologist Karen Emmorey turns to sign language – a language which unfolds in space – for new light on this old problem. For centuries, as she reminds us, the prejudices held against the deaf were also held against their preferred mode of communication. She argues vigorously not just for sign language as a 'proper' language, but one that confers special cognitive powers on its users. In 'Language and space', she draws attention to some fascinating experiments on spatial cognition in signers and non-signers. Accustomed to interpreting visual displays from alternate perspectives, signers generally outperform non-signers on spatial tasks such as the mental rotation of objects. While signers may not inhabit a thoughtworld all their own, their minds – and brains – do show important differences from those of non-signers.

Architects are notably adept at three-dimensional mental 'vision'. Our third author, Daniel Libeskind, has conceived some of the most remarkable and celebrated public spaces of our time, including the Jewish Museum in Berlin, the Imperial War Museum North in Manchester and the buildings chosen for the

World Trade Center site in New York City. His chapter is part memoir, part manifesto. Recalling the origins of some of his most famous designs, he makes it clear that, for him, a failure to engage imaginatively with space has contributed to a malaise in architecture. While other intellectual enterprises, from astrophysics to genetics, from economics to cybernetics, ceaselessly develop radical new approaches to a changing world, architecture – framed by tradition and bound by convention – struggles to be contemporary. 'Architectural space' is a call for reinvention. A building must be 'not a metaphor but a transformation'. It must be at once 'a text' and an 'enclosure of nuts and bolts' – suffused with human memories, sensitive to human needs, inviting to the senses as well as the intellect.

Libeskind's spaces have taken physical form in bricks and mortar, glass and steel, titanium and plaster. Char Davies' spaces, by contrast, exist only in the virtual world defined by computers. As she explains in 'Virtual space', Davies is an artist who constructs immersive virtual reality environments. She charts her personal trajectory from painting to early exploration of the digital medium as a means of artistic expression, going on to describe her two landmark pieces, *Osmose* and *Ephémère*. Where others have used virtual reality technology to simulate familiar worlds, or imaginary worlds explored in familiar (often violent) ways, Davies has created utterly new and beautiful worlds, through which, with the aid of a specially designed helmet and harness, one floats. Direction is controlled through breathing. If one gazes long enough at the quasi-biological objects encountered, they metamorphose. She finds her credo in the French philosopher Gaston Bachelard's *Poetics of Space*: 'By changing space, by leaving the space of one's usual sensibilities, one enters into communication with a space that is psychically innovating . . . For we do not change place, we change our nature.'

From these human-scale spaces we ascend to truly global perspectives on space. In 'Mapping space', the historian Lisa Jardine conjures one of the great eras of global exploration and global mapping, the Renaissance. To be sure, familiar events such as Columbus' discovery of the New World and Copernicus' de-centring of the Earth matter for understanding changes in Renaissance Europe's sense of its literal and metaphorical place in the grand scheme. But too often the role of Europe's relations with its Eastern neighbours is neglected. Jardine's chapter helps to restore the balance. Looking in particular at maps woven into the tapestries that travelled with the Habsburg court in the first half of the sixteenth century, she shows how these apparently neutral representations of the Earth's surface advanced major artistic and imperial agendas of the day. For Jardine, the pivotal contest was that between the Habsburg and Ottoman empires. Out of it, she argues, came not just redrawn boundary lines but – thanks to the manoeuvrings of canny tapestry dealers, eager to see the rival emperors outspend each other on the latest design – the aesthetic so much esteemed by later admirers of Renaissance art.

The need for a longer perspective on East–West relations requires no argument after the defining event of 2001, the terrorist attacks on 11 September (these lectures were given between January and March). Poignantly, the political journalist Neal Ascherson's reflections on 'International space' are organised around the notion of absences, gaps, holes. He draws our attention to the 'spaces between the spaces' which have often eluded the mapmakers. There have, he notes, been literal no-man's-lands: territories that, either by agreement (Antarctica) or accident (the fabled European 'Discrepancy'), have belonged to no nation or empire. But he also considers space as what opens after nations and empires crumble away; or as what, despite native occupants, can be declared in need of tenanting. Metaphors and international space have a long history. Ascherson recalls how natural it seemed in the nineteenth century, in the age of nationalism in politics and the cell theory in biology, to describe nations as cells. For us, who face the distinctive challenges of the twenty-first century, he offers a different metaphor: space as an air pocket hollowed out from within the foundations of political–economic life – space for thinking, feeling and speaking authentically.

'Space' often refers, of course, to extraterrestial space; and that is where this volume concludes. From the Renaissance onward, voyagers have worked to communicate to those who stayed behind what it was like to be in those new and hitherto unknown places. In 'Exploring space', the astronaut and astronomer Jeffrey Hoffman takes up this venerable task. He dwells especially on the human side of space travel. After all, as he points out, only a small number of us have so far journeyed on spacecraft. Perhaps space tourism will, as he suggests, prove a realistic prospect in the near future, so that more can enjoy firsthand impressions. Until then, we have his vivid testimonial to what it is like to ascend into orbit, live and work (in his case, fixing the Hubble Space Telescope), and return. He insists that being in space is just one way to experience it, and not always the best one. For Hoffman, the machines used in space exploration, from telescopes to robots, are our indispensable allies. We should welcome the exploration-by-proxy they offer as enlarging our experience as well as our knowledge of space.

Such knowledge has grown as much through new theories as new data. In 'Outer space', the astrophysicist John Barrow looks at theoretical debates about the ultimate nature of space and the universe. He touches on a number of classic questions. Does space exist over and above the physical objects in the universe (space as 'absolute')? Or is space nothing but the sum of relations between objects (space as 'relative')? How, after Einstein, should we think about space and time, and space and matter? Other questions considered are of more recent vintage, such as whether our expanding universe is accelerating or decelerating, and with what consequences. Most provocatively, Barrow argues that the three spatial dimensions and one time dimension that define our universe are a pre-requisite for the evolution of life. A universe with more than one time dimension, for instance, would not have allowed for information gleaned from the environment to inform future behaviour, making Darwinian evolution impossible. It is, Barrow intimates, as if our universe was designed to enable the emergence of beings intelligent enough to pose questions about its origins and nature.

There ends this particular space odyssey. We referred earlier to Kubrick's *2001* as a source of inspiration. The escalating perspectives of the volume, from the microscale scurryings of molecules in the synapses of a conscious brain to the macroscale rushings away of the edges of the universe, equally call to mind another, perhaps less well-known film: Charles and Ray Eames' *Powers of Ten* (1977). The Eames let their camera move ten times further away every ten seconds. We hope this volume, in its rather different medium, achieves something of the same effect, of making what is familiar about space newly interesting and what is unfamiliar newly relevant.

1 Inner space

SUSAN GREENFIELD

Consciousness: the inner space of the brain

Surely, the ultimate question is, not just for scientists, but also for any human being: 'How does the brain generate consciousness?' Consciousness, to date, has eluded formal definition. Perhaps the easiest would be 'the experience that you are going to lose this evening when you go to sleep, or when an anaesthetist bears down upon you'. Perhaps, for the time being, we could work with the informal idea that it is your own personal world – a world that only you can experience at first hand. If we are to approach, as scientists, how the brain generates this unique inner state, then we must assume that consciousness is generated by, for and of the brain. And yet, any scientific explanation must include the quintessential feature of consciousness: subjectivity.

Before we go any further, we should state what we are *not* going to attempt to do in this chapter. We will leave to one side the issue of how the physical matter of the brain generates the subjective state. We are not, in other words, going into the glamorous water into wine conversion problem of consciousness. Therein lies madness. Philosophers talk themselves round and round that question. What I would like to do instead is look for a correlation of consciousness. I am a pragmatic scientist, and I want to ask questions about consciousness on which we might make some progress. So what we are going to try and do is find whether there is something in the brain, something physical, some brain state that will match up with, or correlate with, different types of consciousness, different ways you feel.

How can we make progress? I would like to concentrate on finding a *neural correlation* of consciousness, one that is both necessary and sufficient. Let us

Space: In Science, Art and Society, edited by F. Penz, G. Radick and R. Howell.
Published by Cambridge University Press. © Darwin College 2004.

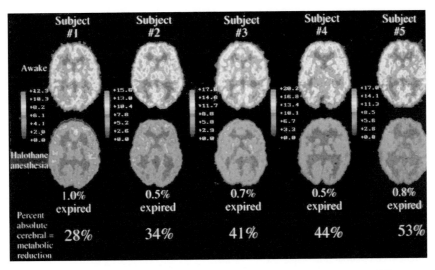

FIGURE 1.1 Effects of loss of consciousness with analysis of activity in different brain regions.

first therefore draw up a list of what we will require of the brain, if it is going to generate different types of consciousness.

One idea is that each brain region has its own set function – that the brain is really a collection of mini-brains. With this idea we would simply enter onto an infinite regress, however, where we miniaturised the problem, but did not solve it. Instead, we know that the brain functions rather like an orchestra, where different instruments of course each play a different part, or a bit like a stew or some complex food, where each ingredient makes its own contribution. With the advent of scanning techniques, this holistic organisation has become clear. Consider, for example, what such techniques have revealed about language and the brain. In one test, the subject was asked to view words passively, listen to words, speak words or generate verbs – all fairly subtle. Even a single 'function' like language, however, is, in terms of the brain, an umbrella for many different processes. Scans show that the brain divides up different aspects of the task to constellations of different brain regions. The important message here is that there is no single brain area lighting up for language, and there is certainly, even for aspects of language, no single brain area that is active. Instead, these regions are, again, working like instruments in an orchestra or like ingredients of some complex dish. They operate all together – the whole is somehow more than the sum of its parts.

This holistic organisation certainly applies to consciousness. If you look at brain imaging again and give subjects anaesthesia, destroying their consciousness, then you can see that there is no single area that shuts down (see Fig. 1.1.). There is no one area of the brain that has just stopped. There is no centre for consciousness.

If there is no centre for consciousness, where is it? It must arise from the participation of non-committed, non-specialised brain cells, or groups of brain cells that, for one moment, enter into some kind of state or configuration that allows for the generation of consciousness. A central factor in the model I shall now describe is that the degree of consciousness therefore is determined by the *size* of an assembly of neurons – *the greater the* size *of the assembly, the higher the degree of consciousness.*

Degrees of consciousness

Since an assembly will have to vary dynamically, we need to look at the factors that will influence its formation, and therefore determine your degree of consciousness at any one moment. Perhaps the most obvious issue is the degree of connections within the brain in the first place. Let me clarify what is at stake by introducing another concept. One source of resistance to this model is the common assumption that consciousness must be all or nothing. A little reflection shows us otherwise, however. Consider the consciousness of non-human animals. Is a dog conscious? If so, what is the difference between it and us? And how does that give us a clue as to what enables our brains to generate different degrees of consciousness?

More controversially, let us extend the riddle to that of the foetus. It is still a commonly held belief that the foetus is not conscious. But if it is not conscious, when does it become conscious? At the time of birth? Fine, but when is a baby born? Some babies are born prematurely, and they are conscious. You would not, for two months, just ignore the baby in the incubator in the hospital and say, 'Ah, forty weeks are up, it is going to be conscious today. Now we can go and visit.' You would be even less likely to do it after the birth: 'Ah, a few weeks have gone now, it is coming up to six months from birth, it might be conscious.' Or is it the manner of birth – squidging down the birth canal? That is tough on babies born by Caesarean section because they will never be

FIGURE 1.2 Shows that you often only see one thing at a time.

conscious. Clearly the manner of birth and the timing of birth, because they are so variable nowadays, cannot be the trigger for consciousness.

I think the big problem here, and one that stops us developing the idea, is normally that we think of consciousness as all or nothing. I myself defined it as 'the thing you are going to lose tonight'. But what if I was wrong? What if instead of consciousness being all or none, consciousness grew as brains grew? What if, therefore, it gradually developed? So a foetus would be conscious, but not *as* conscious as a child, and a baby conscious but not *as* conscious as an adult, and a cat conscious but not *as* conscious as a primate, and a monkey conscious but not *as* conscious as a human. If consciousness grows as brains grow, that raises an interesting issue. As an adult human being, we are more conscious at some times than at other times. If you think about it, we talk about 'raising' our consciousness or 'deepening' our consciousness – it does not matter which way you go, you can go up or down; consciousness is then something that is variable. If that is the case, then science finally has a purchase on the problem – because, instead of looking for some magic brain region or gene or chemical, we can look for something that varies in degree, something we may be able to measure. We can look for something conceivably that ebbs and flows within your brain, something that changes in size within your brain. Now, what could that be?

Let us look at the properties of consciousness, therefore. We have seen there is no special brain region for it. It is spatially multiple, with many brain regions contributing to it. But you only have one consciousness at any one time. I would like to think that you only see one thing at one time. Even when what you see is a very complex pattern, you will still see it as a single pattern. In Fig. 1.2, there is both a young girl and an old crone – which is the true representation? Both are valid, but looking at it one way negates, for a moment, the other. You can only 'perceive' one at a time, rather than 'see'.

As a final item on our list, consider that we are always conscious of something. When we become very sophisticated of course we have an inner hope or fear or dream or thought or fantasy: you can close your eyes and have a trigger of consciousness occurring internally. In the simplest form we have momentary states triggered by the changing input from the sensory world.

A metaphor for consciousness

Let me suggest a metaphor to account for how these properties might be accommodated in the brain. Imagine a stone falling in a puddle. When a stone is thrown, just for a moment, it generates ripples that are highly transient, and that are vastly bigger in their extent than the size of the stone itself; and those ripples can vary enormously according to the size of the stone, the height from which it is thrown, the force with which it is thrown and the degree of competition from other stones coming in (see Fig. 1.3). All these different factors will determine the extent of the ripples.

What I am suggesting is that in the brain you do have the equivalent of puddles: hard-wired little circuits of brain cells, as we have seen, riddling your adult brain, that are sometimes accessed, sometimes not. The equivalent of throwing the stone would be, for example, me seeing my husband. Nerve impulses from my eyes would then go through certain parts of my brain and start activating

FIGURE 1.3 Metaphor for a transient neuronal assembly in the brain.

the circuitry that is related purely to experiences with my husband. Still I would not be conscious of him. What would happen then? How could we now get ripples occurring in the brain? Amazingly enough, you have something very special in your brain: not just circuits, but chemical fountains. These chemical fountains actually emanate from primitive parts of your brain and access the cortex and many other areas. The cortex is the outer folded layer of the cerebrum. These are the chemicals that are targeted, for example, by Ecstasy or by Prozac, and these are the chemicals that vary during sleep–wakefulness and during high arousal: they fulfil a very special function, in that they put brain cells on red alert.

So, imagine I see my husband. That is the equivalent of throwing the stone. It activates a hub of hard-wired circuitry, established, in the example of my husband, over long experience of married life. If that is coincidental with a group of brain cells being sprayed upon by a fountain of chemicals related to arousal, then the process would predispose those adjacent cells to be corralled up just for that moment; and just for that moment a very active hub will then activate a much, much bigger group of cells; and that big group of cells will determine the extent of my consciousness at that particular moment. That is the model.

A correlate of consciousness?

An increasing body of experimental and other data suggests the stone-and-ripple model is correct. Some of these data have come from Arnivam Grinvald's laboratory at the Weizmann Institute in Israel. I was very fortunate in being able to visit Grinvald's lab and see his important experiments at first hand. He showed me experiments – not on humans, because the research is invasive – using optical dyes that register the voltage of brain cells. Arnivam showed that, even to a flash of light, there is indeed the neuron equivalent of ripples. These ripples – in this particular experiment to a simple flash of light – extend over 10 million neurons, and they extend very quickly, in about 230 milliseconds, less than a quarter of a second. So this means that in your brain you can have tens, even hundreds, of millions of brain cells corralled up into a highly transient working assembly in less than a quarter of a second, just like a ripple. That, in my view, is the best place to look if we are trying to find out about consciousness.

You have up to 100 000 connections onto any one brain cell. If you were to count them at one a second in the outer layer of the brain, in the cortex alone, it would take you 32 million years. If you wanted to work out the permutations and combinations, it would exceed the particles in the universe.

Connections are really important, because they work at a level mid-way between genes and larger brain regions. Why are they so important? Let us go back again to see where they fit in. The number of genes you have is about 30 000. Even if one makes the really unlikely assumption that every single gene in your body accounted for a brain connection – even if that was the case – here you have about 10^{15} brain connections (that is 10 followed by 15 zeros), so you would be out by 10^{10}. You just do not have enough genes to determine your brain connectivity. People who hope one day to manipulate their genes so that they are good at housekeeping, or being witty or not being shy – all these other things that people fondly hope that they can start targeting with molecular biology – should, I think, bear this number in mind. There is far more to your brain than your genes.

I am not saying for one moment that genes are not important, and I am not saying that if a gene goes wrong you will not have some kind of terrible malfunction – of course these things can happen: but there is far more above and beyond the single genes that is really important. We are talking here, of course, about nurture, not just nature. The most marvellous thing about humans

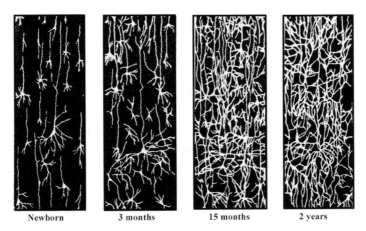

Newborn   3 months   15 months   2 years

FIGURE 1.4 Development of the connections in the first two years of life.

beings and other animals is that as we grow our brain connections grow as well. So, although we are born with pretty much all the brain cells we will ever have, it is the growth of the connections after birth which accounts for the growth of our brains (see Fig. 1.4).

Why is this post-birth growth so important? The answer, and the reason I have been emphasising that genes do not determine brain connections, is that, if brain connections are growing as you are growing, then your brain will mirror what happens to you. Even if you are a clone (that is to say, even if you are an identical twin) you will have a unique configuration of brain cell connections that will shape your reactions to things and will mirror your experiences; so you will see the whole world in terms of your personalised brain cell connections.

Brain plasticity

Let us look at some evidence for this. London taxi drivers, as you may know, are masters at remembering. Every single working day they have to remember how to get from one place to another – and not only the configurations of the streets; they also have to remember the one-way systems and how best to navigate round the streets of London. In one particularly fascinating study, scientists scanned the brains of London taxi drivers and compared them with scans of other people of a similar age. Surprisingly enough, they found that the hippocampus was enlarged in taxi drivers compared to people of a similar age. Now, could it be that people with an enlarged hippocampus are disposed to become taxi drivers? No, because it was found that the longer they had been plying their trade, the more marked this structural difference was. It was what they were doing, their daily activities, that had physically changed their brains.

Your brain will relate to whatever you do. Consider another very simple example: in another study, human subjects were asked to practise five-finger piano exercises. With physical practice, even within five days, there was an enormous enhancement in areas of the brain relating to the digits, just by doing five-finger piano exercises. More remarkable still is that there was a comparable change in brain territory when people were not practising the piano, but were *imagining* they were practising. In real terms you can see that brain territory is reflecting even mental processes, and it is physically measurable.

Nowadays few subscribe to the silly distinction between mental activity and brain activity, as if airy-fairy thoughts were something that floated free, beamed in from Planet Zog. Everything that happens to you, everything you are thinking, has some kind of physical basis rooted in your physical brain. What we are realising now is how exquisitely sensitive the brain is to your experiences and what you do, and therefore how it makes you the individual you are.

Even into old age, one's brain remains continuously 'plastic'; that is to say, it is constantly dynamic, it is constantly evolving and changing, mirroring whatever happens to you. Sadly, what can be created over a lifetime can also be destroyed. Old age often brings with it a reduction in brain connectivity, and consequent losses in inner life and outward functioning. I would like to stress that dementia (which is a name for the confusion and memory loss which characterise Alzheimer's disease) is *not* a natural consequence of ageing; however, when it does strike, it is because the branches with which brain cells make contact with other cells have atrophied. Neurally as well as mentally, therefore, if one becomes senile, it is almost like becoming a child again. As the brain connections are dismantled, you reverse development; and where, in childhood, the world means more and more, so this time the world means less and less. 'Means less': that is to say, you cannot see things in terms of other things any more, because the connections are no longer there.

Mind versus consciousness

When we reflect on the mind of a child, or a senile adult, it is natural to ask: what is a mind? We are clear what brains are, so why do people now talk about minds as opposed to brains? I myself do not subscribe to the idea that the mind is some non-material alternative to the biological squalor that scientists work with. I would like to suggest to you that 'mind' has a very clear physical basis in the brain. We have seen that we are born with pretty much all our brain cells. The growth of the connections between cells accounts for the growth of the brain after birth. These connections, as we have seen, reflect your experience, and in turn they influence your further perception, so you see the world in terms of what has happened to you.

I think this is what the 'mind' is: it is no more and no less than the person-alisation of the brain. You are born, in the words of the great William James, into a world that is a 'booming, buzzing confusion', where you judge the world, and objects in it, as to how sweet things are, how fast, how cold, how hot, how loud, how bright. You evaluate it in terms of its pure sensory qualities. As you get older, these sweet, bright, noisy, loud, fast, cold, hot things acquire labels; they become objects or people or processes or phenomena. They have labels, and then they have memories and associations attached to them, so gradually you can no longer deconstruct the world (unless you are some brilliant artist) in terms of colours and noises and abstract shapes; instead you see it with a meaning, a meaning that is special to you. That is how it continues to occur, and, as we have seen, the connections that sub-serve these processes remain highly dynamic. So, as you go through life, the world acquires a highly personalised significance, built up by 'hard-wired' circuits in the brain.

Although we have this mind rooted in personalised circuits in the brain and we therefore see the world in a certain way, this organisation is not always accessed. Let us consider 'blowing the mind'. Sadly, the patient with dementia is losing their mind on a permanent basis, but, amazingly, some people pay money to lose their minds or 'let themselves go' at raves. The very word Ecstasy means 'to stand outside of yourself'. I think phrases like 'lose your mind', 'blow your mind', 'out of your mind', 'let yourself go' are exactly what we are talking about – although of course, one is still conscious. Conversely, tonight, when you lose your consciousness, I imagine you are not expecting to lose your mind.

The most basic form of consciousness

Let us look further at the relationship between mindfulness and consciousness. Suppose that, in order for consciousness to be mindful at all, the cells active in the brain must include a neuronal assembly of a certain minimum size. We can then ask what would happen if you had just one, abnormally small assembly, so that, though you were conscious, your mind was not operational, or perhaps did not even exist.

Let us just think what kind of consciousness you might have. One advantage of this model is that there are several different scenarios which might result in an abnormally small assembly of brain cells. For example, if the connectivity was modest; or if the epicentre was weak or only weakly activated, as if by a tiny little pebble laid very gently on the surface of the water; or if the production of chemicals malfunctioned; or, indeed, if there was competition from new rapidly forming assemblies – all the different factors could give you different but similarly mind-less types of consciousness.

Modest connectivity occurs in childhood, as we have seen. What do we know about children? Very young children live in the moment. They are not engaged in vast learning and memory tasks. The world does not 'mean' much to them; they judge the world, literally, at face value, on how fast and cold and sweet and so on. They have a literally sensational time, judging the world not in terms of associations but by the impact on their senses at that very moment.

What if the centre of the assembly is only weakly activated? One particular example is one that you are most likely to experience tonight, and that is a dream. A dream could be a small assembly because, in your sleeping state, you do not have your senses working very heavily, so they are therefore not able to recruit a very large assembly and therefore you are dependent on the residual, spontaneous activity of brain cells. Hence, in dreaming, the world seems to be highly emotional, not very logical; you have ruptures in your logic, as with schizophrenic states. We know also that children have a much greater pre-disposition to dreaming than adults do.

Schizophrenia appears to be a consequence of impaired chemical fountaining, in particular of the neurotransmitter dopamine. The connections are all there, the senses are working, but the chemical regulation of the assemblies has gone awry. Again, the similarities between the realities of the dreamer, the child and the schizophrenic are well known. A dream has roughly the same 'mental status' as a psychotic episode. A proverb is as hard to interpret for the literally-minded schizophrenic as for a child.

It seems that all these different states and more – playing a fast-paced sport, dancing at the rave – have one thing in common. They are all characterised by living in the moment, by having a strong emphasis on the senses, and by not placing a great premium on anything you have learned or remembered; that is, not using your mind. The mind is not accessed in any of these states. In schizophrenia, dreaming and childhood, individuals are the passive recipient of their senses, they are out of control, they are not using logic, and they are experiencing strong emotions: an abandonment (literally), or sheer terror, that cannot be constrained by the checks and balances of the sobering adult mind.

Let us summarise: childhood, dreaming, schizophrenia, fast-paced sports, and, dare I say it, raves are all examples of activities and conditions where the mind is not being accessed. It is not being accessed for different reasons: lack of connectivity (childhood), lack of strong sensory stimulation (dreams), an imbalance with those cascading chemicals (schizophrenia) or a degree of competition from other stimulations (fast-paced sports). Indeed the latter would also apply to raves, where, for good measure, people are flooding their brains with a drug that would deliberately interfere with those cascading chemicals.

The converse of pleasure

At one extreme, there is mind-free, pleasurable experience, and a small-to-nonexistent neuronal assembly. Let us now consider the opposite extreme, associated with a large neuronal assembly. At this extreme, the experienced world is grey and remote, where you feel cut off from other people – where, instead of the senses imploding in on you, as they do in dreaming or fast-paced sports or childhood or in schizophrenia as well, you feel numb and remote, your emotions are very turned down: you perhaps feel nothing at all or you do not think you feel anything.

This description captures the features of clinical depression: the outside world is remote, the senses are under-stimulated and there is a continuity of thought, even a persistent thought. In such conditions people do not feel pleasure. It is not that they feel desperately sad; they just feel nothing. They suffer from what is clinically called 'anhedonia', meaning literally 'a lack of pleasure'.

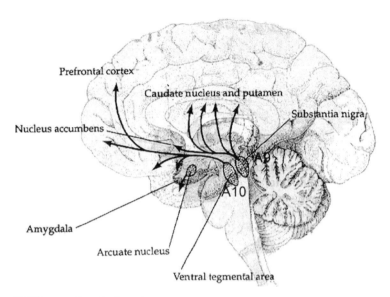

FIGURE 1.5 Gross brain regions.

A depressed person may have the sun on their face or the grass between their bare toes but they do not feel that sensual pleasure that I would like to think we all feel to some measure, and certainly children feel a great deal. They are completely cut off from it.

In clinical depression, then, we have an imbalance of the fountains or cascades of the modulating chemicals and therefore a lack of pleasure and a lack of emotion. What I would like to suggest is that the greater the neuronal assembly, the fewer emotions one feels at any one time. High emotional states, as in childhood or during dreaming or in schizophrenia, are associated with small assemblies. The emotions therefore must be the basic form of consciousness, not learning and memory.

From phenomenology to physiology

Now that we have developed the model, we can begin to test it, to see whether it helps make sense of other observations about mental states and brain states. Consider a subjective experience every single one of us has known: pain. Now most people think of pain as rather obvious, as something that surely we would all feel in the same way if, for example, we put our hands over a flame; but nothing could be further from the truth.

First of all, we know pain is expressed as other associations: pricking, stabbing, burning, chilling. We know that it can vary, interestingly enough, throughout the day. There are some particularly sadistic experiments where volunteers had electric shocks through their teeth and they had to report when they felt the pain. Amazingly, if that happens, you find that throughout the day your so-called 'pain threshold' (when you report that the pain is particularly hard) varies. However, the conduction velocity of your pain fibres does not change, so a factor in the brain is changing, a feature in its chemical landscape. Something transient is changing if you, as the same individual, do not experience the same pain depending on what time of day it is. On the model developed here, an excellent candidate for the feature is the dynamic neuronal assembly. A number of observations about pain become intelligible via the assembly model.

We know that the more people anticipate pain, the more they perceive it as painful; and I would suggest that that is because there is a build-up of the neuron connections. If you are anticipating pain, more and more assemblies are being recruited before you feel it. Incidentally, pain threshold changes throughout the day are due to the modulatory chemical fountains changing. We know that phantom limb pain is felt by people with severed limbs: they do not possess the limb, but they feel as though it is still present. That is because there is a so-called neuronal matrix – which I would call an 'assembly' – corresponding to the severed limb. This could actually be stimulated by the lack of input from externally derived signals.

Pain is absent in dreams – and that, I have suggested, is a small assembly state. Similarly, morphine, which is a very strong analgesic, gives you a dream-like euphoria: the way morphine works is via a natural opiate which acts to make the brain cell assemblies less efficient at being corralled up – therefore people will often say they feel the pain but it no longer 'matters', it is no longer significant to them. Finally, with schizophrenia, which I have suggested is a small assembly state, people, it has been shown, have a higher threshold for pain. For people who are depressed, it is the opposite: they feel pain more. Finally, anaesthetics – which have always proved a puzzle to try and understand because chemically there are many different types of anaesthetics – could work by depressing the activity, so that in the end they gradually reduce the size of an assembly to such a small one that you do not have appreciable consciousness. If that was the case, then you would expect that, as the assembly was shrinking, you would go through the small assembly state and have some kind of delirium, some madness, or, dare I say it, some kind of euphoria or pleasure – a rather odd idea if you are going under an anaesthetic! Nowadays such experiences would be unlikely because anaesthetics are so efficient that they work very quickly – but in the past people would actually have 'ether frolics' or take nitrous oxide at fun fairs; even ketamine, which is an anaesthetic in high doses, is a drug of abuse in low doses. So there is a link there, paradoxically – between slow loss of consciousness and pleasure.

Conclusion

We can speak about these factors 'bilingually', in terms of mental states or in terms of brain states. We can talk about the 'intensity' of our senses, and that is the degree to which your neurons are active. We can talk about 'significance', and that can be the presence of pre-existing associations. We can talk about how 'aroused' you are, and that is the availability of chemical fountains. I have not talked about pre-disposition or mood but that can be other chemicals which also put cells on red alert, like hormones. Finally, we can talk about 'distraction', which bilingually one could refer to as the formation of competing assemblies. So what I have done here is to relate things that you feel to things that could be happening in the brain, and thus to make a start on building up some kind of match. I think in the future we could image these brain cell assemblies and manipulate those factors differentially, making predictions as to the type of consciousness someone might have according to how big their assembly was, or vice versa. Whether or not this model is correct, its real strength is that it can be tested – if not now, then in the future.

Finally, we must remember that the brain is an integral part of the body, that your central nervous system, your hormones and your immune systems are all inter-linked; otherwise there would be biological anarchy. So these brain cell assemblies, which I am suggesting are related to consciousness, are merely an index of consciousness. If you put one in a dish, you would not have con-sciousness. Somehow it causes a readout to the rest of the body and some-how the rest of the body signals back via chemicals that influence the size of the assembly and hence consciousness. One candidate group of chemicals are the peptides, which can also function as hormones and could co-ordinate the immune, endocrine and nervous systems.

In conclusion, we will never be able to get inside someone else's brain. Inner space will remain private. On the other hand, I think that science is starting to make a contribution by being a little more modest. By actually matching up physical brain states with what people are feeling, we may acquire some insight into why people take drugs, what happiness is, and, perhaps most puzzling of all, why people go to raves or go bungee jumping.

FURTHER READING

Carter, Rita, *Consciousness*, London: Weidenfeld Nicolson Illustrated, 2002.

Chalmers, David, *The Conscious Mind: In Search of a Fundamental Theory*, Oxford: Oxford University Press, 1997.

Damasio, Antonio, *The Feeling of What Happens: Body, Emotion and the Making of Consciousness*, London: Vintage, 2000.

Dennett, Daniel C., *Kinds of Minds: Towards an Understanding of Consciousness*, London: Phoenix, 1997.

Edelman, Gerald M., *Bright Air, Brilliant Fire: On the Matter of the Mind*, London: BasicBooks, 1992.

Greenfield, Susan, *The Human Brain: A Guided Tour*, London: Phoenix, 1998.

Greenfield, Susan, *The Private Life of the Brain*, London: Penguin, 2002.

McGinn, Colin, *The Mysterious Flame: Conscious Minds in a Material World*, London: BasicBooks, 2000.

Rose, Steven (ed.), *From Brains to Consciousness? Essays on the New Sciences of the Mind*, London: Penguin, 1999.

Searle, John R., *The Mystery of Consciousness*, London: Granta Books, 1998.

2 Language and space

KAREN EMMOREY

Another title for this chapter might be 'Languages in space' because the chapter primarily explores the nature of signed languages – languages perceived by eye, rather than by ear, and produced with manual rather than with vocal articulations. Specifically, we examine what the study of signed languages can tell us about the nature of human language, about spatial cognition, and about the brain. The focus of our exploration is on how the visual–gestural nature of signed languages affects (a) the structure of language, (b) the spatial cognitive abilities of signers, and (c) the neural organisation of language.

First, however, we briefly explore the domain of 'language and space' in spoken languages. *Spatial language* refers to how we talk about what we see. Languages differ with respect to how spatial distinctions observed in the world are categorised by words. For example, English and Korean use spatial terms that categorise spatial concepts in strikingly different ways. English makes a distinction between actions resulting in containment (*put in*) and support (*put on*). In contrast, Korean makes a cross-cutting distinction between tight-fit relationships (*kkita* refers to a peg put *into* a hole or to a magnet placed *on* a surface) and loose-fit relationships (*nehta* refers to blocks loosely placed *in* a pan). The Korean verb *kkita* refers to actions resulting in a tight-fit relation, regardless of the containment or support relationship. In contrast, English uses two different prepositions, *in* and *on*, to distinguish containment from support. Melissa Bowerman and her colleagues found that very young children are sensitive to these language-specific distinctions in spatial categorisation. Korean children less than two years old pay attention to tight-fit spatial arrangements when they hear *kkita* (regardless of the containment relation between the objects),

Space: In Science, Art and Society, edited by F. Penz, G. Radick and R. Howell.
Published by Cambridge University Press. © Darwin College 2004.

whereas English-learning children attend to containment relationships when they hear *in*, regardless of the tightness-of-fit relationship.

Thus, languages not only categorise spatial concepts differently, but there is evidence that language learners are influenced by the semantic categorisation of space in their input language from a remarkably early age. As soon as Korean- and English-speaking children begin to talk about spatial relationships, they use spatial terms in ways that reflect the spatial categories encoded by their respective languages. These results show that very young children must learn to 'structure space for language'. That is, children must attend to how their language packages spatial information. It appears that children may not learn spatial words simply by mapping these words to prelinguistic concepts of space, such as containment or support. Since Korean has a spatial category ('tight fit') that cross-cuts containment and support, one would predict that Korean-speaking children would acquire the 'tight fit' category later than English-speaking children acquire the spatial words *in* and *on*, which neatly map to the prelinguistic concepts of containment and support. However, there is no delay in acquisition for Korean-speaking children, suggesting either that 'tight fit' and 'loose fit' are also prelinguistic spatial categories or that children must *build* spatial semantic categories based on the language they hear (perhaps in addition to relying on prelinguistic spatial categories).

The fact that languages differ in their spatial semantic categories raises the fascinating question of whether such linguistic differences have an impact on nonlinguistic spatial cognition. This question is the topic of much current debate and research. An example of how linguistic spatial categories might impact spatial thinking comes from studies of the Mayan language Tzeltal, spoken in Chiapas, Mexico. Tzeltal lacks words that express the spatial concepts of 'left', 'right', 'front' and 'back'; instead, absolute co-ordinates, roughly equivalent to 'north' and 'south', are used to describe the location of objects in space. These terms are used not only to describe the location of landmarks in large-scale environments, but also for small-scale locations. For example, a Tzeltal speaker might describe a spoon as 'south' ('uphill') of a plate on a table. As one would expect, given the system of spatial contrasts in their language, Tzeltal speakers are extremely good at dead reckoning, much better than Dutch speakers whose language does not require them to be constantly aware of an external frame of reference. Furthermore, Stephen Levinson and colleagues have shown that Tzeltal speakers and Dutch speakers differentially categorise and encode spatial scenes for memory. For example, if shown an arrow pointing left/east on a table, Dutch speakers remember the arrow as 'pointing left' and will reproduce the arrow pointing to their left if turned around within the room. In contrast, Tzeltal speakers remember the arrow as 'pointing east' and will reproduce the arrow pointing east (not to their left), if turned around in the room and asked to make the arrow point in the same direction. Thus, how people remember spatial information is systematically influenced by the language they speak.

Signed languages are particularly interesting to study with respect to the relationship between language and space because these languages use space itself to express linguistic contrasts. In this chapter, we first explore how the visual and spatial modality of signed languages influences linguistic structure, addressing the question 'How are signed languages different from and/or similar to spoken languages?' Next, we examine how signers and speakers talk about space. What are the consequences of using space itself to talk about space? Third, we explore whether the habitual use of a signed language impacts spatial cognition. Do signers exhibit enhanced visual–spatial skills compared to speakers? Finally, we examine whether signed and spoken languages are represented similarly in the brain. Are the same neural regions recruited for both signing and speaking?

Before we begin, however, it is important to quickly debunk several common myths and misconceptions about sign languages.

Myths and misconceptions about sign languages

Myth number 1: there is a universal sign language

No sign language is shared by deaf people of the world. There are many distinct sign languages that have evolved independently of each other. Just as spoken languages differ in their lexicon, in the types of grammatical rules they contain, and in historical relationships, signed languages also differ along these lines. For example, despite the fact that American Sign Language (ASL) and British Sign Language (BSL) are surrounded by the same spoken language, they are mutually unintelligible. Sign languages are most often named after the country or area in which they are used, for example Mexican Sign Language, Swedish Sign Language, Taiwan Sign Language. It should be noted that the exact number of sign languages in the world is not known. The signed languages that will concern us here are not 'secondary' sign languages, such as the sign system used by Trappist monks or the sign language used by hearing Australian Aborigines during mourning (when silence is required). These systems differ from the primary signed languages of Deaf communities in ways that suggest strong links to an associated spoken language. Specifically, the grammar of these secondary systems reflects the grammar of the surrounding spoken language. In addition, secondary signed systems are not the native language of anyone, i.e. they are not learned during childhood nor do they constitute the primary means of everyday communication for their users.

Myth number 2: sign languages are based on oral languages

American Sign Language has mistakenly been thought to be 'English on the hands'. However, ASL has an independent grammar that is quite different from the grammar of English. For example, ASL allows much freer word order compared to English. English contains tense markers (e.g. *-ed* to express past tense), but ASL (like many languages) does not have tense markers that are part of the morphology of a word; rather tense is often expressed lexically (e.g. by adverbs such as *yesterday*). There are no indigenous signed languages that are simply a transformation of a spoken language to the hands. There are invented systems used in educational settings that manually code spoken language. They are not acquired in the same manner as natural signed languages, and they do not arise spontaneously.

One might ask 'If sign languages are not based on spoken languages, then where did they come from?' However, this question is as difficult to answer as the question 'Where did language come from?' We know very little about the very first spoken or signed languages of the world, but research is beginning to uncover the historical relationships between sign languages, as has been done for spoken languages. For example, the origin of American Sign Language can be traced to the existence of a large community of deaf people in France in the eighteenth century. These people attended the first public school for the deaf, and the sign language that arose within this community is still used in France today. In 1817, Laurent Clerc, a deaf teacher from this French school, along with Thomas Gallaudet, established the first public school for the deaf in the United States. Clerc introduced French Sign Language into the school. The gestural systems and indigenous signs of the American children attending this school mixed with French Sign Language to create a new form that was no longer recognisable as French Sign Language. ASL still retains a historical resemblance to French Sign Language, but both are now distinct languages.

As noted earlier, British Sign Language (BSL) is completely unrelated to ASL. The roots of British Sign Language can be traced at least to the seventeenth century. References to deaf people in England using signs to communicate can be found in documents from the 1600s. The development of deaf education in Britain brought isolated deaf people together into schools in the mid eighteenth century. The lexicon of modern BSL can be traced back to the signs used in these schools. In addition, through the immigration of British deaf people, the signed languages used in New Zealand and in Australia are strongly related to BSL (and are not at all related to ASL).

> Myth number 3: sign languages cannot convey the same subtleties
> and complex meanings that spoken languages can

On the contrary, sign languages are equipped with the same expressive power that is inherent in spoken languages. Sign languages can express complicated and intricate concepts with the same degree of explicitness and eloquence as spoken languages. This particular myth apparently stems from the misconception that there are 'primitive' languages (this label is often applied to oppressed peoples; for example, it was used by turn-of-the-century Westerners to describe African languages).

In addition, poetry and song are forms of artistic expression that are found in both spoken and signed languages. One might ask, 'How are poems and songs expressed in a language without sound?' ASL poetry exhibits a patterning of form that is quite parallel to the poetic sound patterns found in oral languages, but it also contains poetic structure that is intrinsic to its visual–spatial modality. Like spoken poetry and song, ASL can rhyme by manipulating the form of signs. For example, signs can rhyme if they share the same handshape or the same movement, and a poem may contain signs that all share a single handshape or group of handshapes that are formationally similar. Rhythmic structure within a sign poem can be created by manipulating the flow of movement between signs and by rhythmically balancing the two hands. The use of space is a poetic device that is intrinsic to the visual modality. Signs move through space and are clustered and separated within space to produce an additional dimension of structure within a poem. Sign poetry also takes advantage of the visual modality by using 'cinematic' techniques such as zooms, close-ups and visual panning.

> Myth number 4: sign languages are made up of pictorial gestures and are similar to mime

On the contrary, sign languages have an intricate compositional structure in which smaller units (such as words) are combined to create higher-level structures (such as sentences). This compositional structure is found at all linguistic levels: *phonology* (the patterning of sounds in spoken languages), *morphology* (the internal structure of words), *syntax* (the grammatical structure of sentences) and *discourse* (the structure of narratives or conversation). It may seem odd to use the term *phonology* since there is no sound in sign language; however, it turns out that sign languages exhibit systematic variations in form that are unrelated to meaning distinctions. That is, there is a componential level of structure below the level of the morpheme (the smallest meaningful unit) that is akin to phonological structure in spoken languages. This is described in more detail below. Such complex and hierarchical levels of structure are not present in pantomime. Pantomime differs from a linguistic system of signs in other important and systematic ways as well. For example, pantomime is always transparent and iconic, but signs can be opaque and arbitrary. For example, the signs APPLE and SIT (see Fig. 2.1 – signs are notated as English glosses in uppercase) bear little resemblance to the form of an apple or the action of sitting. With pantomime, however, one could mimic holding and eating an apple to represent the concept of 'apple' or mime sitting on an invisible chair to convey the concept of 'sitting'. In addition, the space in which signs are articulated is much more restricted than that available for pantomime – pantomime can involve movement of the entire body, while signing is constrained to a space extending from just below the waist to the top of the head. Such restrictions permit fluent and rapid articulation of conventional symbolic forms using just the hands and face. Finally, the ability to pantomime and the ability to sign can be affected differently by brain damage, indicating different neural systems are involved.

FIGURE 2.1 Illustration of part of the phonological system of American Sign Language. The figure provides examples of minimal pairs in ASL: (A) signs that contrast only in hand configuration, (B) signs that contrast only in place of articulation, and (C) signs that contrast in movement. Illustration from Emmorey, *Language, Cognition and the Brain*; copyright Lawrence Erlbaum Associates; reprinted with permission.

Space and linguistic structure

Signed languages make use of spatial contrasts to express linguistic information at several linguistic levels, from phonology to discourse. We begin by asking 'How can we even talk about spatial contrasts as "phonological" when the system is not sound-based?'

Many linguists have argued that signed languages exhibit a phonological level of structure, despite the fact that signed languages are perceived visually rather than auditorily. In spoken languages, words are constructed out of sounds, which in and of themselves have no meaning. The words *bat* and *pat* differ only in the initial sounds, which have no inherent meanings of their own. Sounds may be combined in various ways to create distinct words: *bad* differs from *dab* only in how the sounds are sequenced. Similarly, signs are constructed out of components that are themselves meaningless and are combined to create morphemes and words.

Signs are composed of four basic phonological elements: handshape, location (place of articulation), movement, and palm orientation. Fig. 2.1 provides an illustration of three *minimal pairs*: signs that are identical except for one component, and if you substitute one component for another, it changes the meaning of the sign. The top figure shows two ASL signs that differ only in handshape. ASL contains over thirty different handshapes, but not all sign languages share the same handshape inventory. For example, the 't' handshape in ASL (the thumb is inserted between the index and middle fingers of a fist) is not used by Danish Sign Language. Chinese Sign Language contains a handshape formed with an open hand with all fingers extended except for the ring finger which is bent – this hand configuration does not occur in ASL. Signs also differ according to where they are made on the body or face. Figure 2.1B shows two signs that differ only in their place of articulation. Spatial locations on the body distinguish minimally between signs, but these locations do not convey meaning. Movement is another contrasting category that distinguishes minimally between signs, as shown in Fig. 2.1C. Finally, signs can differ solely in the orientation of the palm; for example, the sign WANT is produced with a spread hand with the palm up, and FREEZE is produced with the same handshape and movement (bending of the fingers and movement toward the body), but the palm is facing downward. These meaningless phonological elements are combined and sequenced to create lexical signs.

Thus, both signed and spoken languages exhibit a level of structure below the word that can be considered *phonology*. For spoken languages, phonology consists of contrasts between sounds that distinguish words ('b' differs from 'd'); whereas for signed languages, contrasts occur between gestural elements that make up the form of signs. One of these elements is spatial location, i.e. where a sign is articulated on the body. At the phonological level of structure, spatial location does not convey meaning, but is simply an element of sign form.

Like words in all human languages, but unlike gestures, signs belong to lexical categories or basic form classes such as noun, verb, modal verb, adjective, adverb, pronoun and determiner. Sign languages have a lexicon of sign forms and a system for creating new signs in which meaningful elements (morphemes) are combined. However, sign languages differ from spoken languages in the type of combinatorial process that most often creates complex words. For spoken languages, complex words are most often formed by adding prefixes or suffixes to a word stem. In sign languages, complex signs are created by nonconcatenative processes, i.e., processes that do not add a morpheme to the beginning or end of the sign. To form a morphologically complex sign, various movement patterns are superimposed onto a sign stem. For example, Fig. 2.2 provides illustrations of the ASL verb GIVE with a number of distinct movement patterns that indicate distinct types of actions over time (different temporal aspects).

For both spoken and signed languages, lexical rules are governed by constraints on their combination and on their application to particular forms. For example, Fig. 2.2A illustrates the citation form of the verb GIVE. This is the form of the verb produced in isolation, without any context. Figure 2.2B shows the verb GIVE with the durational inflection, a movement pattern meaning 'continually over time'; and Fig. 2.2C shows GIVE with the exhaustive inflection, meaning 'to each and all'. In the form of GIVE shown in Fig. 2.2D, the durational inflection applies after the exhaustive inflection to yield a form meaning 'give to each in turn, over a long time'. Such a verb could be used to describe someone at Hallowe'en giving out candy to children, again and again, throughout the evening. In contrast, if the durational inflection applies prior to the exhaustive, as in Fig. 2.2E, the meaning of the verb is 'give continuously to each in turn' which could be used to describe a teacher passing out several papers to each student in a class. Finally, the durational inflection can apply recursively, before and after the exhaustive, as in Fig. 2.2F. A verb inflected in

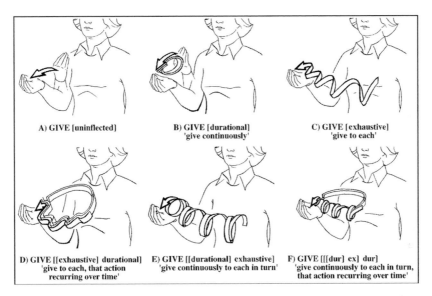

FIGURE 2.2 Illustration of morphological processes in American Sign Language. The meanings of the verbal inflections are given in brackets. The durational inflection indicates an action occurs for a long time. The exhaustive inflection means roughly 'to each and all'. Illustrations copyright Ursula Bellugi, The Salk Institute.

this way could be used to describe a teacher passing out several papers to each student, and this action occurs throughout the day (e.g., for each class). Thus, the meaning of a verb is dependent upon how morphological inflections are applied to the verb.

English does not mark temporal aspect, such as 'continually over time', on verbs, but other spoken languages do. Although signed and spoken languages both mark the same type of morphological information on verbs, the way it is encoded is very different for the two language types. Spoken languages tend to prefer affixation, i.e., prefixes and suffixes added to a word. Although signed languages have some prefixes and suffixes, they are rare. Signed languages tend to encode morphological information by superimposing movement patterns onto a verb stem, as shown in Fig. 2.2. This preference for non-linear, simultaneous morphology appears to arise from the visual–spatial modality. The visual system is very good at perceiving information simultaneously, whereas the auditory system is very good at making fine temporal distinctions within a linear string. Furthermore, the vocal articulators are very quick, and speakers can produce roughly 12–15 distinct sounds per second. In

FIGURE 2.3 Illustration of the use of space at the level of the sentence in ASL. The figure illustrates the association of nouns with locations in space (indicated by subscripts), and the movements of a verb between these loci (indicated by an arrow). The spatial endpoints of the verb indicate the subject and object of the sentence.

contrast, the hands are much slower articulators, and signers produce roughly 7–11 phonological distinctions per second. The articulatory–perceptual requirements of sign conspire to create a preference for simultaneous, non-linear word structure.

Finally, at the level of the sentence, signed languages appear universally to use space for referential functions. For example, in ASL and other signed languages, nouns (or noun phrases) can be associated with locations in signing space. This association can be established by 'indexing' or pointing to a location in space after producing a lexical sign, as shown in Fig. 2.3. Once a referent, such as 'dog' or 'cat', has been associated with a spatial location, the signer may then refer back to that referent by using a pronominal sign directed toward that location. In addition, verbs can move with respect to these locations to indicate subject and object relationships. In Fig. 2.3, DOG has been associated with a point on the signer's left and CAT with a location on the signer's right. The verb BITE moves from left to right, and means 'The dog bites the cat.' If the verb were to move from right to left, it would mean 'The cat bites the dog.' The use of space to express co-reference and semantic–grammatical relations is a unique resource afforded by the visual nature of sign language.

The study of space and linguistic structure in signed languages tells us something about the nature of human language in general. There are enormous differences between the tongue and hand as articulators, as well as between the visual and auditory systems. Nonetheless, we find striking parallels between the structure of spoken and signed languages, even at the level of phonology, traditionally defined as the *sound* patterns of language. At the level of word structure, we find that sign languages encode the same kinds of semantic concepts as spoken languages (e.g., temporal distinctions expressed by verb inflections). The difference in modality is seen in how these distinctions are linguistically encoded. Sign languages prefer simultaneous morphology in which inflections (movement patterns) are superimposed on a verb, whereas spoken languages prefer linear affixation with prefixes and suffixes. At the level of syntax we find that spoken and signed languages obey the same universal constraints on sentence structure. However, signed languages use spatial contrasts to convey grammatical distinctions. Research with signed languages indicates that the properties that have been found to be universal for spoken languages do not arise just because these languages are spoken. That is, they do not arise because of the characteristics of speech, since they are observed in signed languages as well.

The use of space to talk about space

In ASL, as well as in other signed languages, signing space can function topographically to represent spatial relations among objects. *Signing space* is the term used for the three-dimensional space in front of the signer, extending from the waist to the forehead, where signs can be articulated. Signers schematise this space to represent physical space, as well as to represent abstract conceptual structure. For example, a signer can associate abstract concepts (e.g., states of matter) with spatial locations in an arrangement that depicts a structural relationship between those concepts (e.g. the order of physical change from solid to liquid to gaseous matter).

For spatial expressions in ASL, there is a schematic correspondence between the location of the hands in signing space and the position of physical objects in the world. When describing spatial relations in ASL, the identity of each object is first indicated by a lexical sign (e.g., HOUSE, BIKE). The location of the objects, their orientation, and their spatial relation vis-a-vis one another is indicated by where the appropriate *classifier* signs are articulated. Figure 2.4 provides a simple illustration of an ASL locative sentence that could be translated as 'The bike is near the house.'

Locative expressions such as the one shown in Fig. 2.4 involve *classifier constructions*. A classifier construction is a complex predicate in which the handshape specifies information about object type. This handshape is sometimes referred to as a classifier because it 'classifies' or specifies certain properties of

| HOUSE | whole-entity CL + *loc* | BIKE | whole-entity CL + *loc* |

FIGURE 2.4 An illustration of a simple spatial description in ASL, using classifier constructions. An English translation would be 'The bike is near the house.' *Whole-entity CL* refers to the type of classifier handshape morpheme and +*loc* refers to the position movement morpheme (a short downward movement) that means 'to be located'. Illustration from Emmorey, *Language, Cognition and the Brain*; copyright Lawrence Erlbaum Associates; reprinted with permission.

the object it refers to. For example, in Fig. 2.4 the hooked 5 handshape specifies a large bulky object (such as a house or box), and the 3 handshape (thumb, middle, and index fingers extended) refers to vehicles (such as a bicycle, car or ship). The first and third signs of Fig. 2.4 (HOUSE and BIKE) are nouns that refer to the ground and figure objects, respectively. The figure object is the located (or moveable) object, and the ground object is the reference (unmoving) object. The sign for BIKE is made with the right hand, while the left hand holds the classifier sign referring to the house. This temporal ordering of ground before figure may be an effect of the visual–spatial modality of sign languages. For example, to present a scene visually by drawing a picture, the ground object tends to be drawn first, and then the figure is located with respect to the ground. Thus, if drawing a picture of a bike next to a house, most people draw the house first.

Crucially, the spatial relationship expressed by the classifier construction in Fig. 2.4 is not encoded by a separate word as it would be in English with the preposition *near*. Although ASL has prepositions such as NEAR, ON or IN, signers prefer to use classifier constructions when describing spatial relationships. Rather than encoding spatial information with prepositions, such information is conveyed by classifier constructions in which there is an analogue mapping between where the hands are placed in signing space and the locations of objects being described. Classifier constructions appear to be universal to sign languages and exhibit properties unique to these languages that arise from the visual–spatial modality.

Spatial language in signed languages differs dramatically from spoken languages because, for the latter, spatial information is conveyed categorically by a small set of closed-class forms, such as prepositions. Spoken languages do not have a way of altering the form of a preposition to mean 'upward and to the left' vs 'upward and slightly to the left' vs 'upward and to the far left'. In ASL, such spatial information is indicated in an analogue manner by where the hands are located in signing space. This is not to say that there are no linguistic constraints on where or how the hands are positioned in signing space. However, the gradient and analogue descriptions of spatial locations that can be produced by signers stand in stark contrast to the categorical nature of spatial descriptions produced by speakers.

The impact of sign language use on visual–spatial cognition

What are some of the implications of this spatialisation of language for non-linguistic visual–spatial abilities? Does the visual-spatial processing that is required for the production and comprehension of spatial descriptions in ASL have any impact on non-linguistic spatial processing? Recent investigations have shown that experience with sign language appears to enhance or alter performance within several cognitive domains. The effects appear to be due to linguistic experience, rather than to deafness, because hearing people who learned ASL as their first language from their deaf parents exhibit the same patterns of performance as deaf ASL signers.

Motion processing

Signers categorise motion patterns differently from non-signers, grouping linguistically significant motions together. When deaf native signers and hearing non-signers were asked to make three-way comparisons of motion displays, they differed in their motion similarity judgments (i.e., judging which two motion displays were most similar). In particular, movement repetition and cyclicity were much more salient for the deaf signers, and repetition and cyclicity are both linguistically distinctive in ASL (see Fig. 2.2). Signers were more likely to categorise these motion patterns as similar, compared to non-signers. These results suggest that the acquisition of American Sign Language can modify the perceptual categorisation of motion patterns. ASL signers categorise certain motion patterns differently from non-signers for whom the patterns carry no linguistic information.

In addition, signers have a heightened sensitivity to certain perceptual qualities of non-linguistic motion. Specifically, they appear to be better able to distinguish between transitional and purposeful motion, compared to non-signers. For example, in one study signers and non-signers were asked to draw Chinese characters that they saw 'written in the air'. They only saw a point-light display of the stroke pattern because the character was drawn in the dark with a small LED attached to the index finger of the drawer. Because neither the signers nor the non-signers were familiar with Chinese, no linguistic knowledge could be used to determine the underlying stroke pattern of the character. Thus, the movement patterns were all non-linguistic for these subjects. ASL signers were significantly better at segmenting the continuous moving-light image into discrete movement strokes. Deaf signers were better able to distinguish between the transition and stroke components of the movement and were less likely to include transitional movements in their drawings of the character, compared to the hearing non-signers. This enhanced ability to analyse movement may arise from the processing requirements of ASL. Signers must separate out meaningful from transitional (non-meaningful) movement during sign perception. This linguistic processing skill may make signers more sensitive to the distinction between purposeful movement and transitional movement.

Finally, signers recruit a different set of brain areas when they perceive motion in the periphery of vision. Specifically, both hearing and deaf signers show much more left-hemisphere involvement when attending to motion in the periphery, compared to non-signers. The increased role of the left hemisphere in motion processing may arise from the temporal coincidence of motion perception and the acquisition of ASL. That is, the acquisition of ASL requires the child to make linguistically significant distinctions based on movement. If the left hemisphere plays an important role in acquiring ASL (as it does for spoken languages), then the left hemisphere may come to mediate the perception of linguistically relevant motion *and* non-linguistic movement patterns. These results suggest that the acquisition of a signed language can alter the brain areas responsible for certain aspects of motion perception.

Face processing

The face carries both linguistic and emotional information for ASL signers. Both hearing and deaf people use their face in the same way to convey emotional information – these expressions (e.g., happy, sad, angry) are universal. However,

ASL signers also use facial expressions to convey linguistic contrasts. Different facial expressions indicate different syntactic structures, such as wh-questions (e.g., what, where, when questions), relative clauses, conditionals, adverbials and topics. These facial expressions differ from emotional expressions in their scope and timing and in the facial muscles that are used. Grammatical facial expressions have a clear onset and offset, and are coordinated with specific parts of the signed sentence. For example, conditional ('if') clauses are signaled by raised eyebrows, a head tilt to the side, and a slight forward movement of the shoulders. The onset and offset of these features signal the beginning and end of the conditional clause. Linguistic facial expressions also co-occur with verbs to convey adverbial information. For example, the facial expression called *mm* (lips pressed together and protruded) indicates an action done effortlessly, whereas the facial expression called *th* (tongue protrudes between the teeth) means 'awkwardly' or 'carelessly'. These two facial expressions accompanying the same verb (e.g. DRIVE) convey quite different meanings ('drive effortlessly' or 'drive carelessly').

Thus, ASL signers have a very different perceptual and cognitive experience with human faces – both linguistic and emotional facial expressions must be quickly discriminated and identified during language processing. Results from several experiments indicate that both hearing and deaf ASL signers exhibit enhanced performance on face-processing tasks, and we have begun to uncover the underlying mechanisms for this enhancement. Specifically, experience with ASL enhances face-processing skills which are most relevant to interpreting subtle changes in facial feature configurations which must be generalised over individual faces – exactly those skills necessary to identify and discriminate among linguistic facial expressions. ASL signers have an enhanced ability to discriminate among very similar faces and to recognise subtle changes in individual facial features. For example, signers exhibit superior performance on the Benton Faces Test, a test that assesses the ability to distinguish among similar faces that differ in lighting and profile. In addition, signers are significantly more accurate than non-signers in discriminating between faces that are identical except for a change in a single facial feature. However, signers and non-signers do not differ in their ability to recognise individual faces from memory or in their gestalt (holistic) face-processing abilities. To identify and categorise ASL facial expressions, signers do not need to recognise the person, and the gestalt aspects of the face (e.g., the shape of the face) remain unaffected by variations in linguistic facial expression. Sign language use appears to enhance only those face-processing skills that must be generalised across unique individuals and that are relevant to interpreting subtle differences in local feature configurations.

Mental imagery

ASL signers exhibit a superior ability to generate and transform mental images. Signers have faster reaction times for tasks that involve mental rotation – for example, deciding whether two objects are the same or mirror images, regardless of orientation. Signers are also faster at tasks that require them to generate an image (e.g., of a letter). Image generation may occur frequently during ASL discourse because, under certain conditions, signers articulate verbs as if they imagine referents as present in the space around them. For example, the verb GIVE (see Fig. 2.2A) would be directed upward if the imagined referent were a very tall person (e.g. a basketball player) and downward if the imagined referent were a small child. Furthermore, classifier constructions often require relatively precise representation of visual–spatial relations, and this explicit encoding may require the generation of visual mental images. Finally, mental rotation may be involved in understanding certain types of spatial descriptions when signers are face-to-face. When understanding the description of a spatial scene from a specific viewpoint (e.g., the entrance to a room), the addressee has mentally to reverse the spatial arrays created by the signer. For example, a spatial location established on the right of the person signing (and thus on the left of the addressee) must be understood as on the right in the scene being described. Experience with such mental reversals may enhance non-linguistic mental rotation abilities. Thus, the habitual use of various imagery skills during sign language processing appears to facilitate the generation of mental imagery within non-linguistic domains.

It is important to point out, however, that there are several visual–spatial cognitive processes that do not appear to be influenced by sign language use. For example, signers and non-signers do not differ with respect to motion discrimination thresholds (the threshold at which objects are detected as moving together), face recognition ability, and the maintenance of visual images in short-term memory. In addition deaf and hearing subjects have an equal ability to distinguish between rapidly presented visual flashes (visual discrimination). There is also little difference between deaf and hearing children in their

Table 2.1 *Summary of the impact of sign language use on visual–spatial cognition*

Affected domains	Unaffected domains
• Mental rotation and image generation	• Visuo-constructive abilities
• Face discrimination and processing of facial features	• Face recognition and gestalt face processing
• Motion categorisation and parsing	• Visual discrimination thresholds
• Neural organisation for motion processing	• Motion detection and contrast sensitivity thresholds
• Memory for spatial locations	• Memory for visual images

From Emmorey, *Language, Cognition and the Brain*; copyright Lawrence Erlbaum and Associates, reprinted with permission.

visual–spatial constructive abilities, i.e., drawing, copying and block construction. A summary of the basic findings is presented in Table 2.1.

Habitual use of a signed language does not create a general enhancement of visual–spatial cognitive processes; rather, there appears to be a selective effect on certain processes argued to be involved in sign language production and comprehension. In particular, motion analysis and categorisation, spatial memory, mental image transformations, and facial feature discrimination are enhanced or altered by experience with sign language. However, much more research is needed to determine the extent and nature of these patterns of performance. For example, are there any low-level perceptual processes that are affected? Are there processes that might actually be impaired due to interference from sign language processing? How similar does a non-linguistic visual–spatial process have to be to the sign language process in order to be affected? Answers to these questions will help us to understand what factors determine whether a given visual–spatial ability will be affected by exposure to sign language.

The neural organisation for sign language

Decades of research indicate that the left hemisphere of the brain is dominant for language such that linguistic abilities are more impaired by damage to the left hemisphere than to the right. The study of signed languages provides special insight into the basis for this cerebral asymmetry. For example, the left hemisphere may be specialised for processing linguistic information or for more general functions on which language depends, such as rapid auditory processing or the execution of complex motor movements. Research with deaf signers who have suffered unilateral brain injury indicates that the left hemisphere is dominant for sign language, despite the modality difference. Left hemisphere damage produces sign language aphasias, while damage to the right hemisphere does not. Sign aphasias are similar to those observed with spoken language. For example, signers with left hemisphere damage may produce effortful signing with little grammatical morphology. Lesions to a different area within the left hemisphere produce fluent signing with many paraphasias (phonological or semantic errors that create nonsense signs). Signers with left hemisphere damage also have much poorer sign comprehension compared to those who suffer damage to the right hemisphere.

This pattern of linguistic deficits does not appear to be simply a function of deficits in general spatial cognitive ability. Right-hemisphere damaged signers exhibit much more severe impairments of visual–spatial abilities such as perceiving spatial orientation, apprehending perspective within a drawing, or interpreting spatial configurations, compared to signers with left hemisphere damage. Furthermore, the difference in linguistic impairment between left- and right-lesioned signers is not a function of group differences in age at time of test, age of the onset of deafness, or age of first exposure to ASL. None of these variables correlate with linguistic impairment (as measured by the ASL version of the Boston Diagnostic Aphasia Exam).

Data from studies with deaf signers without brain injury also indicate left hemispheric specialisation for sign language processing. Various techniques can be used to measure neural function in normal signers. For example, positron emission tomography (PET) measures regional cerebral blood flow and relies on the link between blood flow and neuronal metabolism. When neurons become activated, they require an immediate increase in glucose and oxygen, and thus blood flow is increased to regions surrounding these neurons. Blood flow within the brain is measured by injecting a short-lived radioactive tracer into the blood stream (the amount of radiation is two times the natural background radiation one receives annually living in Denver, Colorado). The tracer preferentially flows toward areas of high blood flow. Thus, neuronal activation can be imaged by recording emissions from the tracer to localise those regions of the brain that exhibit higher rates of cellular metabolism and increased blood flow.

Using the PET technique, researchers have found left-hemisphere activation when deaf signers are asked to produce signed words or sentences. The activated regions (left inferior frontal cortex) correspond to the same areas that are engaged during speech articulation by hearing subjects. Visual–spatial areas within the right hemisphere do not show significant activation during signing (or speaking). In addition, left-hemisphere activation is found regardless of whether signers produce signs with their dominant right hand (controlled by the left hemisphere), with their left hand (controlled by the right hemisphere), or with both hands. This result indicates that the left hemisphere specialisation for sign language production is not simply due to left hemisphere control of the dominant right hand in signing.

Another neuroimaging technique is called Functional Magnetic Resonance Imaging or fMRI. Traditionally, Magnetic Resonance Imaging (MRI) has been used to provide high-resolution images of internal anatomy, from knees to hearts to brains. Within recent years, however, the technique has been adapted to provide *functional* images of brain activity. It turns out that the magnetic resonance (MR) signal is sensitive to changes in the metabolic state of the brain. Small changes in the MR signal arise from changes in the magnetic state of hemoglobin which carries oxygen in the blood. To be specific, de-oxygenated hemoglobin is paramagnetic (responds to an induced magnetic field), while oxygenated hemoglobin is not. Thus, the fMRI technique measures the changes in blood oxygenation that accompany neural activity.

Several fMRI studies with deaf and hearing native signers have found neural activation within left hemisphere structures that are classically linked to spoken language comprehension (e.g., Broca's and Wernicke's areas). In addition, some studies have found more activation within the right hemisphere when signers comprehend ASL sentences than when speakers comprehend spoken sentences. However, other studies have found similar regions of activation within the right hemisphere for both spoken and signed language comprehension. The right hemisphere appears to play a larger role in language comprehension than in production for both types of languages. However, whether the right hemisphere plays a slightly greater role in signed language comprehension is currently a topic of research.

The study of deaf and hearing signers can provide unique insight into the neurobiological bases for language. By contrasting *form* (visual–gestural vs auditory–oral) and *function* (e.g., communication), generalisations about what determines the neural organisation of language can emerge. Research on the neural bases of sign language indicates that left hemispheric specialisation for language is not based on hearing or speech. Given that sign language relies primarily on spatial information rather than on rapidly changing temporal information to encode linguistic distinctions, left hemisphere dominance for language does not arise from a general competence for processing fast temporal changes. Furthermore, left hemisphere damage can spare the execution of complex motor movements that are non-linguistic (e.g., pantomime or non-representational movements) but impair the production of sign language. Thus, the disruption to sign language is not due to a general disruption of motor control. The results from signers with unilateral brain lesions and from neuroimaging studies with normal subjects suggest that the basis for the left hemispheric specialisation for language lies in the nature of linguistic systems rather than in the sensory characteristics of the linguistic signal or in the motor aspects of language production.

Summary and conclusions

This chapter briefly illustrates what can be learned about human language, cognition, and the brain by studying signed languages and the deaf people who use them. The linguistic results thus far reveal substantial similarities between signed and spoken languages, but this is only a starting point. The similarities provide a strong basis for comparison and serve to highlight universal properties of human language. Linguistic investigations are also beginning to uncover clear distinctions between signed and spoken languages, and these distinctions reveal how language modality can affect linguistic preferences (e.g., a preference for simultaneous morphology over linear affixation). Signed languages also clearly differ from spoken languages in their ability to use signing space to represent both spatial and non-spatial information.

For signed languages, the linguistic representation of spatial information systematically differs from that in spoken languages in ways that reflect the structural characteristics of scene parsing during visual perception. For example, Leonard Talmy has discovered that spoken languages combine information about motion path with information about the ground object. In the sentence, *The bike went past the house*, the prepositional phrase *past the house* expresses both the path of the bike (it went *past*) and the ground object (*the house*). This combination does not reflect how we actually perceive the visual scene because we observe the figure object (the bike) as moving along a path. In contrast, signed languages represent visual–spatial information in a manner consistent with visual scene parsing (i.e., how we interpret objects moving through space). In signed languages, the figure object is combined with path motion because the classifier handshape representing the figure object (the bike) is combined with movement to indicate path. For example, to express 'The bike went past the house', the classifier handshape representing the bike in Fig. 2.2 would move past the classifier handshape representing the house. The classifier system of signed languages appears to be much closer to the structural characteristics of visual parsing than the closed-class (e.g., prepositional) systems of spoken languages. This more parallel structure between linguistic expression and visual scene parsing can also be seen in the ability of signers to produce more gradient and analogue descriptions of spatial locations, as described earlier in the chapter.

Because signed languages rely on visual–spatial cognitive processes, they provide a unique tool for investigating the relation between linguistic and non-linguistic domains of cognition. Parallel studies are difficult to conduct with spoken language because non-linguistic auditory processing cannot be studied in the absence of experience with speech. Evidence from studies with signers (deaf and hearing) and non-signers indicates that the language one uses can enhance certain cognitive processes through practice. Through habitual use within the language domain, cognitive processes can be faster (as with image generation), more fine-tuned (as with face discrimination and aspects of motion processing), or more adept at coding certain types of information (as with memory for spatial sequences). These effects of language use on cognitive behaviour go beyond the 'thinking for speaking' hypothesis put forth by Dan Slobin. Slobin's hypothesis is that the nature of one's language (in particular the grammatical categories of one's language) affects cognitive processes *at the moment of speaking*. For example, if a language has the grammatical category number (e.g., singular, dual, plural), then a speaker must attend to number while speaking, whereas a speaker of a language that does not obligatorily encode number does not need to pay attention to the number of objects. Results with users of American Sign Language, as well as the recent work by Stephen Levinson with Tzeltal speakers, indicate that the language one uses can influence cognitive processes even when speaking/signing is not required.

Finally, with the accessibility of new brain imaging techniques, there is currently an explosion of studies investigating the neural systems underlying sign language production and comprehension. Thus far, the data indicate nearly identical neural structures supporting sign and speech. This result argues against the hypothesis that the co-evolution of language and the neuroanatomical mechanisms of speech production are what led to the left hemisphere specialisation for language. Rather, it may be that neural structures within the left hemisphere are particularly well suited to interpreting and representing linguistic systems, regardless of the biology of language production and perception. The critical question, of course, is *why* are these neural structures well suited for language, or, put another way, what is it about linguistic systems that causes them to be left lateralised? These questions remain unanswered, but the study of signed languages provides a tool by teasing apart those aspects of linguistic systems that are fundamental and inherent to the system from those aspects that can be affected by the nature of language perception (visual vs auditory) or by the nature of language production (manual vs oral). The study of signed languages provides an unusual technique for exploring how the brain tolerates and adapts to variation in biology.

FURTHER READING

Bavelier, D., Brozinksy, C., Tomann, A., Mitchell, T., Neville, H. and Liu, G., 'Impact of early deafness and early exposure to sign language on the cerebral organization for motion processing', *Journal of Neuroscience*, **21** (2001), 8931–42.

Bowerman, M., 'How to structure space for language', in *Language and Space*, ed. P. Bloom, M. Peterson, L. Nadel and M. Garrett, Cambridge, MA: MIT Press, 1996.

Corina, D. P., 'Aphasia in users of signed languages', in *Aphasia in Atypical Populations*, ed. P. Coppens, Y. Lebrun and A. Basso, pp. 261–310, Mahwah, NJ: Lawrence Erlbaum Associates, 1998.

Emmorey, K., *Language, Cognition, and the Brain: Insights from Sign Language Research*, Mahwah, NJ: Lawrence Erlbaum and Associates, 2002.

Emmorey, K. (ed.), *Perspectives on Classifier Constructions in Signed Languages*, Mahwah, NJ: Lawrence Erlbaum and Associates, 2003.

Levinson, S. C., Kita, S., Haun, D. and Rasch, B. H., 'Returning the tables: language affects spatial reasoning', *Cognition*, **84** (2002), 155–88.

Neville, H., Bavelier, D., Corina, D., Rauschecker, J., Karni, A., Lalwani, A., Braun, A., Clark, V., Jezzard, P. and Turner, R., 'Cerebral organization for language in deaf and hearing subjects: biological constraints and effects of experience', *Proceedings of the National Academy of Science*, **95** (1998), 922–9.

Slobin, D., 'From "thought and language" to "thinking for speaking"', in *Rethinking Linguistic Relativity*, ed. J. J. Gumperz and S. C. Levinson, Cambridge: Cambridge University Press, 1996, pp. 70–96.

Talmy, L., *Toward a Cognitive Semantics* Vol. I: *Concept Structuring Systems*, Cambridge, MA: MIT Press, 2000.

3 Architectural space

DANIEL LIBESKIND

The space of language is rather different from the space of architecture. I hope that I will be able to transmit something to you that is not only about the explicit external condition of architectural space. Everybody has experienced space, and, in a way, describing that experience could be analogous to St Augustine's famous dilemma of describing time. Everybody knows it, but when it comes to really analysing it, one gets lost. So it is with space, which under scrutiny suddenly becomes political space, social space, economic space, abstract space, subversive space, pathological space or potential space.

To investigate the essential role of architectural space and how it is generated, an old parable from the classical philosophers comes to mind: that architecture is the home of man but at the same time it is something that is built and constructed by architects, by builders. The goal of architecture on the one hand, and the mechanics of making a building on the other, are two different things. I try to navigate between these two poles. Architecture is not only a matter of construction, not only about putting together some homogeneous system of elementary modules. It is about fusing irreconcilable dimensions, analysable in geometric terms and practical terms, but also in dimensions that are ineffable and worldly.

As you will know, there are different kinds of architects with different notions of architectural space. Certainly, I am not one of those who believe that architecture is to be found in 'the box', or in the decadent notion of detailing the box to death and increasing its costs. I think of architectural space as an adventure, an adventure that has an obscure genesis and an open history. This state is, to quote Voltaire (Parodying Leibniz), the best of all possible worlds. But it is still only one of many possibilities.

Space: In Science, Art and Society, edited by F. Penz, G. Radick and R. Howell.
Published by Cambridge University Press. © Darwin College 2004.

FIGURE 3.1 Reading machine © Hélène Binet.

The dramatic transformations of architectural space are mysterious and cannot be explained simply by the passage of time. The sudden changes of human desire are things that I am very interested in. Architectural space is an ephemeral space, a space that has as much to do with faith in what is not there as the substantiation of what could be there. All of which makes architectural space as complex as it is fascinating.

Some time ago I meditated on the problem of what constituted the dimensions of architecture, besides the obvious dimensions of length, height and width. Architecture is indeed three-dimensional because it is a cross-section of the cultural world. So I investigated the idea that the three dimensions of architecture are about reading, writing and memory. Reading architecture is not reading text, but reading in the sense of communicating and deciphering texts that communicate something in no clearly explicit language. Writing architecture is not writing literary text, but writing in the sense of inscribing ourselves in the book of possibilities which include unknown configurations of relationships, names, places, people, dates, and the light which reflects and refracts with architecture. The third dimension, the dimension of memory, brings architecture into reality.

I was fortunate to be given the opportunity to produce three machines for the 1985 Venice Biennale competition, a project that was awarded the Lion of

FIGURE 3.2 Writing machine © Hélène Binet.

FIGURE 3.3 Memory machine © Hélène Binet.

Venice (see Figs. 3.1, 3.2 and 3.3). I thought of these machines as necessary in order to suggest a new an urban *idea*, and not only a piece of architecture. I wrote a number of texts and created three machines: one for reading architecture, one for writing architecture and one for remembering architecture. The reading machine is a kind of wooden medieval machine, in whatever country or whatever space. This is a simple mechanical machine with gears and wedges. The permanent rotation of the same text provides circularity without redundancy and the geometric location of centres and peripheries.

The second machine is the writing machine, a more complex mechanism. It has a quadrilateral rotation in a completely unpredictable pattern. This is a machine based around seven cubes and seven handles and seven gears that rotate in unpredictable, ever impossible configurations. The totality of its four-sidedness includes an empty plane of light, a plane for saints, a plane of emblems and a plane of the lines of the city of Palmanova. The turning of the cubes is effectively an unstable technology. The writing machine destabilises the technological way of making because it is geared to a movement that is at any one time synchronic, more like dreams which gear and unlock themselves.

The third machine, the memory machine, is a very particular one because it is a theatre of the mind, a theatre of discourse. It is the gadgetry, so to speak, of the permanence of renaissance, the rediscovery through papers, levers, puppetry, drawers, all sorts of devices, which bring architecture into some form of coherence as an art.

I produced a special plan of Berlin while I was working on a competition for the development of Potsdamer Platz (see Figs. 3.4 and 3.5). Special, because I think that any plan for a city is not just a plan but more like a mechanism, a mechanism that itself produces gods! The city and its transformations refer to that transcendent dimension which is incarnated not only in the lines of the streets and the urban spaces that we know, but in other energetic connections: the true matrix of the city. I called this plan the Muse-Matrix. I used the nine muses, because the city is about the lyric, about the tragic, about the poetic, about rhetoric, about the narrative, astronomy, dance, about music and economy. It is as much about the graveyard as it is about the schoolhouse; as much about the tavern as it is about the hospital. The city is a kind of labyrinth that reformulates our perception. Architectural space is not always coincident with the space that we see in the plans of cities.

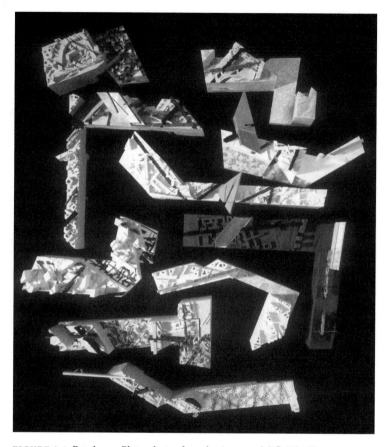

FIGURE 3.4 Potsdamer Platz: photo of puzzle pieces model © Udo Hesse.

In this project for the Potsdamer Platz, I invited some of my friends to specify, as precisely as possible, the hour and second of a singular date in the twentieth century which they deemed to be of the greatest significance. I can show you only a very few, but I received many – there were a lot of interesting people involved. Heine Müller responded with a cryptic set of numbers (see Fig. 3.6). One particularly interesting piece came from Jacques Derrida, who actually drew the plan of a city in a very daring way by giving it a kind of black erased mark in the middle (see Fig. 3.7).

Now, thinking about the figure of a city through events that have to do with dates and anniversaries of some sort does not always coincide with a spatial matrix. The fragments themselves have to be discovered. All of these conditions

FIGURE 3.5 Potsdamer Platz: photo of puzzle pieces model © Udo Hesse.

turn into a kind of rebus, a different form of making a city which is not made from ready-made pieces to be fitted together in a homogenous system, but rather fusing actual dimensions which do not have any explicit connection. I imagined the Potsdamer Platz, the figure of Berlin, in a completely new way (see Fig. 3.8). It is, nevertheless, also a completely old figure of Berlin. What is this old figure of Berlin? Where is one to find it? It is not something you find in the plans of Potsdamer Platz from 1933 to 1939; it is that star-like burst of human possibility.

I thought of some of the great city plans – like Rome's. If you lived in Rome back in Roman times, you would have thought that it was eternal and would last forever. Even walking through the scraps of ruins of the Roman Forum now, you can feel the confidence that these people had that their horizon would eternally be repeated in the circularity of their own inscribed memories. But it came to an end. Castles in Europe have the same aura of total reliability. Similarly, if we go to Chicago, London, New York, we confidently look at the

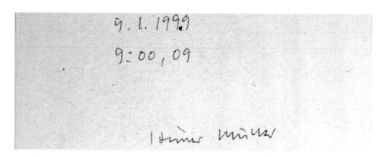

FIGURE 3.6 Potsdamer Platz: photo of fax from Heine Müller.

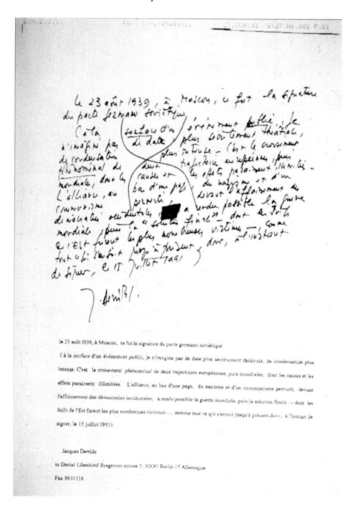

FIGURE 3.7 Potsdamer Platz: photo of fax from Derrida.

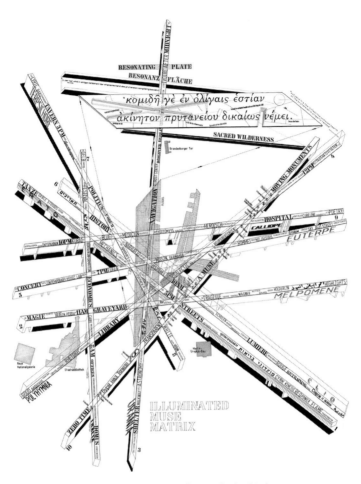

FIGURE 3.8 Potsdamer Platz: drawing © Daniel Libeskind.

new world; and yet there is no doubt that it won't be like that forever, in that way. This new matrix for Berlin produces different forces of density, relationship and spatiality. Drawing and apprehending a city in this way might seem very obscure, especially when the drawing done is in a completely different meditative sphere. My drawings are not, however, drawings of impossible buildings or hypothetical utopias, but rather are ones that have their own logic of reading, writing and remembering.

Recently, I won a competition in Switzerland for a very large shopping centre, a different type of project from ones I normally do. It is about 150 000

FIGURE 3.9 Theatrum Mundi: photo © Daniel Libeskind.

square metres – a whole city with housing, shopping, hotels, leisure and parks.
I think I won it on the basis of a new kind of three-dimensional matrix. As an
architectural space, the centre developed from an original and highly unusual
set of drawings that of course then evolved. So what I am really interested in
is not simply taking a ready-made plan of history or a type of history, and pro-
jecting it into the future. I find it far more intriguing to delve into some sort of
a root, transmitting and communicating those rooted elements of the city and
architecture to the crown of what is visible in memorable architectural space.

As a conceptual project, I produced a set of drawings or paintings done as a
project of architecture, the Theatrum Mundi (see Fig. 3.9). I attempted to depict
the Earth as it looks after the events that we have witnessed in the twentieth
century. What is that space of communication across territory that was devastated through ideologies? Has that architectural space physically been emptied
of everything? By emptying, I mean the flattening out of meaning. That architectural space: what does it mean, how does it develop? I produced drawings
that move across a certain *somatic* space because architecture is produced not
just by the head. It is produced in every way and has to be communicated, as

a public event. These are drawings that explore different types of experiments where there is a connection between the wildest intuition and something that can finally become an object.

Most of these drawings are done by hand. I do not believe that the manipulations of computers give us more freedom – quite the contrary. Computers are based on the linear development of the perspectival notion of developing ideas to an end. I am interested in something rather looser: a different configuration, which as I have said has to do with reading, with the emblem, with memories, with the unknown names, with rooted addresses, points, beings, souls.

The Felix Nussbaum Haus is a project that I built in Osnabrück, Germany (see Figs. 3.10 and 3.11). It is a museum built for a painter who perished during the Holocaust. I organized the museum in the same way that the Theatrum Mundi drawings were designed – devastated spaces which coincide, collide and are precisely allied around the vectors of his life: between his hometown of Osnabrück, his hopes for Berlin, his studies in Rome, his imprisonment in France, his escape to Brussels, his concealment in Antwerp and then the end station of deportation and death in Auschwitz.

That matrix of connections, fixed and never changing, will always exist. It is a strange eternal constellation that cuts across our own experience. This is a museum connected to history, not in a very obvious way but in a rather raw way. How does a building invoke the dead-ends of a human being's life? I called this project, from the very beginning, the 'Museum without Exit' – in German, 'Museum ohne Ausgang' – because it is not a museum in which you move in a circular route and reaffirm the beauty of circularity. Rather, the building stands as a text, as a voyage into that quadrilateral rotation which does not come back to a uniform set of planes of light. The space goes beyond into a memory that is fragile and vulnerable. Undoubtedly, 'Museum ohne Ausgang' is disturbing, because you come across spaces that do not lead anywhere, and yet they do lead into the walls and across the walls. Of course, we cannot physically go into a wall or through a wall, but we do anyway, through the paintings and through the works of Nussbaum. In this building you can see some of the architectural devices which are used, and not as metaphors. Some people may believe that this kind of architecture is about metaphor, because of a drawing, or a book, or a text, or a vision, or even a measured space. But it is not about a metaphor, it is about transformation. It is about transforming these dimensions into an open realm and relating them to the ultimate destination of our own lives.

FIGURE 3.10 Felix Nussbaum Haus, Osnabrück: photo, view from above
© Bitter&Bredt.

FIGURE 3.11 Felix Nussbaum Haus, Osnabrück: photo, Nussbaum Gang
© Bitter&Bredt.

The Nussbaum Gang is at the heart of the building. It is an extremely narrow space that may look like a corridor, but is not. It is a double-storey space, an acoustical space: the whole centre of the building, although it is on the periphery. I tried to design the narrowest space that could be built under German regulations, and I fought for it and convinced the authorities. The space is less than 2 metres, 1.80 metres, and I still think it is too wide. It is not claustrophobic, but it does give that relationship to the emptiness, the hugeness of the external concrete wall vis-à-vis the old buildings that surround that project. It is an empty wall, the biggest unpainted canvas of Nussbaum. Within the Nussbaum Gang are the paintings that Nussbaum painted while hiding from the Gestapo. I wanted to create a space for those paintings which would allow the visitor to have the kinetic and emotional experience of what it felt like to paint a painting whilst in hiding, in cramped attic rooms, having no distance to what is on the canvas.

Surrounding the museum is a field of sunflowers, perhaps not the usual landscaping for a museum, but the sunflower was Felix Nussbaum's favourite flower and thus I used it. As it happened, the children of Osnabrück have adopted these sunflowers and take care of them while they are flowering.

Of course, in the end, one might ask: what is this building about? It is really about memory, and the fact that there is a continuous present. Nussbaum's paintings were thrown into the ocean of history like a message in a bottle. Only by accident – really by a small accident – did someone find that desperate message in the bottle and bring it back to this location.

The discovery and return of Nussbaum's paintings is somehow a destiny, the destiny I hoped to give to that museum in an obscure and never contemplated way. And yet it is inscribed into objective materials like the world itself.

Many years ago, as another conceptual idea, I tried to make a model of the globe, as a different kind of mapping. This was a conceptual project, an architectural world, a spherical construction that, as it were, ran along rails. As I thought about the project, I came to see it as something that was in fact completely broken. The world is not really as substantial or as solid as it appears. This research evolved into my idea for the Imperial War Museum North of Manchester. I took the very vulnerable crust of the earth and I shattered it. When I reassembled it, in three large shards, it was no longer the perfect globe that we see on CNN or the BBC. It is something that has been transformed,

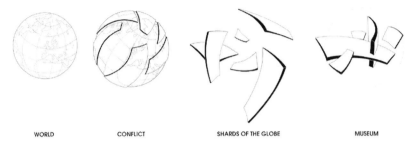

WORLD   CONFLICT   SHARDS OF THE GLOBE   MUSEUM

FIGURE 3.12 Imperial War Museum – North, Manchester: conceptual drawing, shattered globe © Daniel Libeskind.

deformed, abused. Yet, this broken globe is constructed to demonstrate the effect of conflict on people all over the world.

Three transformed shards, representing air, earth and water, were reassembled to create the building (see Fig. 3.12). The air shard serves as the entrance to the museum, rising 65 metres. Right at the top of this dramatic building is an observation deck that looks over the entire landscape of Manchester. The earth shard, which is covered by a double curved roof, houses the temporary exhibit space and the permanent exhibition, as well as the education block. The floor of the earth shard is curved, so that when visitors enter, they are physically 2 metres higher at the center than at the horizon. The water shard is an inverted curve and contains the cafe, restaurant and gift shop that overlook the Peel ship canal. The earth shard has various heights within itself. At some points it is 20 metres high, at other points the roof almost touches the ground. Nevertheless, it has a distinct homogeneity about it. The fourth element is the fire of the exhibits themselves. The exhibition is an interesting mix of 'image totale' and the physical objects of war. Stories are told of the people from the North and illustrated in 20 metre high images. At the same time, within the exhibition spaces are silos that allow groups of up to sixty people to gather and ask presenters to call up objects of all previous wars. From the virtual to the real, from huge images and stories to landmines and gas masks, the visitor will have the opportunity to put themselves into the conflicts that eternally plague the world.

I remember a profound thought of Winston Churchill, who said that the one thing that we can rely on is the permanence of conflict. How is that prophecy to unfold? The Imperial War Museum North is a museum with a mission to communicate to people something about the great effort of cities, the tragedy of cities, the transformation of nations in the seconds, minutes, hours and days of war. I am particularly aware of this, as I am working on a project in Dresden, that beautiful Baroque city which suffered almost total destruction within hours. It is horrifying to see those eternal ruins. Architecture *is* an art that can also be about recovery. It is about the light that it throws and its own re-materialisation.

In all the projects that I have developed, I have attempted to deal with how one combines space which can be read – a text which is not an architectural space as such – with an architectural closure that has bolts, panels, systems of construction. How does one traverse through a material space and make it into architecture? Architecture and memory have been underestimated. I am not talking about the nostalgia for what we have seen, but rather the idea of the contemporary. Think of an old man, sitting in a room next to a window with a fireplace, with a bird, a nightingale in a cage. The man is looking out of the window. Without memory, this would be the ultimate scene of madness, of absurdity, of the irreconcilable. And yet, if you tell the story of the man, it seems completely reasonable to have the nightingale caged, the fireplace behind the man and the window open. Undoubtedly, architectural space has a role to play in the story of memory.

For the competition for the Jewish Museum in Berlin, I made a drawing before the unification of Berlin, before 9 November 1989. It is a drawing of the distorted Star of David over the plan of Berlin; and as I was drawing it, I deliberately drew the lines connecting and crossing through the Berlin Wall (see Fig. 3.13). Of course, historians and others had told me that the Wall would be there forever, so why draw lines that cross through into another country? Why indeed? For the simple reason that when it came to the history of Germans and Jews, the Berlin Wall was invisible. This history is not something very obvious, because it is not about a conventional notion of fatality or tragedy, but rather the total abyss created at the centre of European culture.

The drawings rotated and dissected that star through various constellations of both impossibility and possibility: to gardens; to exile; through displacement; into the realm of social justice; into a hopeful future. These competition drawings for the Jewish museum are difficult to read and yet somehow I feel they

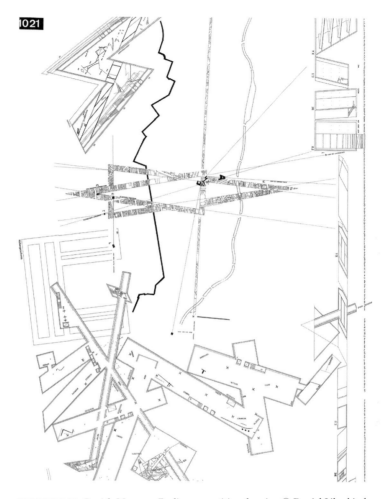

FIGURE 3.13 Jewish Museum, Berlin: competition drawing © Daniel Libeskind.

protected the outcome. Perhaps the jury chose my scheme because they felt that it could never be built!

For this project, I had many, varied, inspirations. I was interested in Walter Benjamin's book *Einbahn Strasse*, the *One Way Street*. How did that text become incorporated in the space of the Museum? Through the apocalypse, over the void, on thirty-six bridges, left and right. And here, the yellow-covered score of Schönberg's *Moses and Aron*, a piece of music which I was very much interested in, not as a metaphor, but as a structure that the museum would make accessible. The final third act of Schönberg's opera is really a conversation, since the lines of music remain empty. I sought to evoke through the space of architecture the completion of that opera. Arnold Schönberg was forced to leave Berlin, retract his hopes and reincarnate himself as Aron Schönberg, leaving behind the silence of the third act. Still another inspiration was the *Gedenkbuch*, the Memory Book, a solemn list of all the deported and murdered Jews from Berlin – names that reach into millions – and my hope was to read into the building every one of those names, date of birth, date of deportation and the presumed date of murder.

The name 'Berlin' itself interested me, not as the abstract name of the city, but as the name of people. I discovered that Jews took on that name because the city offered them the extraordinary opportunity of being free for the first time, free to write books in Hebrew as a secular language, free to join in the spirit of the city.

Can such a heavy building stand on such thin ground? For it does stand on very thin ground, because it is the ground of those names. I thought that the ground of Berlin is not just the solid opacity of space, but is both the air above and the ground below. A few centimetres upward or downward and you are lost in a sort of dream of what was there and what might still be.

The void traverses space at any time; but at the Jewish Museum in Berlin it is physically constructed to be made visible, not as a hypothetical or metaphysical concept, but as something to be touched, seen and crossed. I wanted to build a completely empty space (see Fig. 3.14), in which nothing could ever be shown because it is all about absence. It is meant to be a space of transition and of thought. At the same time, I designed and constructed an upside-down garden: a garden of 49 pillars filled with earth, 48 with the earth of Berlin and standing for the formation of the State of Israel (in 1948), and one filled with the earth of Jerusalem and standing for the city of Berlin. The pillars are absolutely perpendicular to the ground upon which they stand, but the ground is canted in two different directions, so that the visitor feels a little disoriented. The garden is about trees that don't grow vertically. It is about exile, displacement, not only to London, New York, Tel Aviv, but within the space of the city itself.

The Jewish Museum was, by the way, my first building, my first chance to build, and so it did not arise from a vast previous experience. Of course, during my studies I had learned how to make foundations, and concrete, and

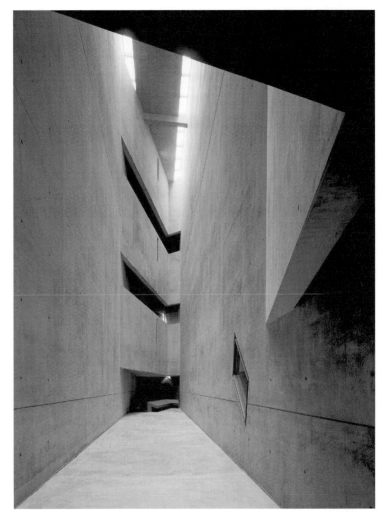

FIGURE 3.14 Jewish Museum, Berlin: 6th void © Bitter & Bredt.

doors and windows. These things are crucial because without getting them right, nothing else is possible. But designing and making this Museum was also about another kind of legibility and writing, of footsteps, and the souls and the hearts of visitors.

Figure 3.15 shows a project entitled 'Chamberworks'. I drew a line, very small, with a very thin pen on a sheet of gray paper. Each line was like a razor

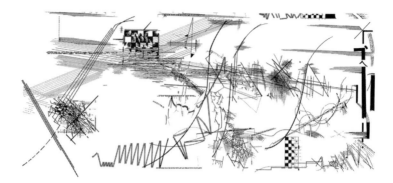

FIGURE 3.15 Chamberworks: drawing, vertical III © Daniel Libeskind.

blade, so that with each line drawn the paper comes off as you draw. I spent a long time doing these drawings, fully expecting that they would have no future. Only later I discovered that buildings have to maximise the potential structure of walls; and these very lines were the key.

I continued the idea of a wall as structure for my project for the extension to the Victoria and Albert Museum in London. I call it the Spiral (see Figs. 3.16 and 3.17). The Spiral is just a single wall, because, I thought, what is a museum? It is, in essence, a wall: you need walls, to separate, to hang, to communicate, to project, to go through. The decentred Spiral is structurally integral because on each wall there are two cuts which hold the structure. Cecil Balmond, my collaborator from the firm Ove Arup, is planning not to construct this building out of rods of steel, just propped up and covered with a light material, but to build it really tectonically, from the bottom up. It will be a solid building, made of concrete, getting lighter and lighter as it carries its own walls within itself. The walls will cantilever out over the Pirelli Court, allowing the visitor to see the spectacular roofs and landscape of London.

Why did I choose a spiral form? As I see it, the Victoria and Albert Museum is about the spiral of knowledge – something referred to by its founder, William Morris. The museum was established by great social thinkers, great artists, craftsmen, and visionaries who knew the ineffable. They proposed such prosaic ideas as including the first restaurant in a museum – the first one in the museum world was at the Victoria and Albert – because workers should be there to eat, and to work and to dream. This was and is a museum dedicated to the

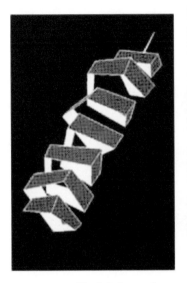

FIGURE 3.16 The Spiral-extension to the Victoria and Albert Museum: drawing, Spiral concept © Daniel Libeskind.

FIGURE 3.17 The Spiral-extension to the Victoria and Albert Museum: photo planning permission model © A. Putler.

tradition of the applied arts, an educational institution to compete with similar developments in Europe.

The walls ascend and open up, creating unprecedented spaces providing for all the diversity of programmes within: interstitial spaces for small video auditoriums, demonstration rooms, huge galleries and mezzanines. Seamless links, six in total, provide crucial circulation and entry to previously cut-off collections in historically sensitive and Grade A listed buildings. The Spiral also opens up the forecourt on Exhibition Road, allowing the visitor to walk fully around the building.

As it happens, I was fortunate enough to construct for an exhibition instal-lation a large-scale model of the Spiral, made of steel and clad in steel, at the Netherlands Architecture Institute (see Fig. 3.18). It was built as a self-supporting structure. During the last few days before the exhibition was to open, the engineers from Ove Arup panicked and decided to put 30 000 more

FIGURE 3.18 Netherlands Architecture Institute: photo, exhibition © Bitter&Bredt.

screws and bolts into the structure! Inside there will be a singular space in which theatres are part of the gallery space connecting to the links, bringing both natural light and artificial light. It is complex, and, by the way, the complexity is an important part. Architecture is not about 'let's get it to the simplest state possible.' Of course, in the end it has to look very simple. And it will.

We have developed a totally new idea for the façade of the Spiral which will be clad in ceramic tiles. Why ceramic? The collection of the Victoria and Albert Museum has an extensive ceramic component; and although it is a traditional material, it is also used on the Space Shuttle. Cecil Balmond and I discovered that the fractile pattern which will determine the tiles is from the Fibonacci series, underpinned by the geometry of the Golden Section. How incredible that the Golden Section should be a part of the dynamic of the non-periodic, non-repetitive geometry! The series will consist of right-angled tiles produced and fabricated, we hope, by an English manufacturer.

The Victoria and Albert Museum has its sceptics, those who believe that architecture is threatened by philistines. They would like to return architectural space to political space, to technical space, to empty space. But I believe that architecture should be left to its own enigmatic and open possibility. London is the most fantastic place, and it deserves, I think, more than just new buildings behind old walls. It needs something which is also on the street, which projects the confidence of its citizens to the outer edges of experience. In architectural space, the public is the most imaginative component, the reader, the writer and its memory. Architecture is always something which is in process and not simply an end result. I think at the end we do wind up at the edge of memory, at the edge of what is memorable and eternal. And that is how I began the competition for the Victoria and Albert, by standing on Exhibition Road and looking upwards. On either side of the door is a figure, and under the figure an inscription: on the left is 'Inspiration' and on the right is 'Knowledge'. Those figures – very beautiful, charming, modest merely standing there – are a testament to the straightforwardness of the Victorians, who were interested not in metaphors, but in metamorphosis.

FURTHER READING

Bunschoten, Raoul, Binet, Helene and Libeskind, Daniel, *A Passage Through Silence and Light*, London: Black Dog Publishers Ltd, 1997.

Dorner, Elke, *Das Judische Museum Berlin*, Berlin: Gebrüder Mann Verlag, 1999.

Freireiss, Kristin (ed.), *Jewish Museum*, Berlin: Ernst & Sohn, 1992.

Levene, Richard C. and Cecilia, Fernando (ed.), *El Croquis: Daniel Libeskind*, Madrid: El Croquis, 1996.

Libeskind, Daniel, *Between Zero and Infinity*, New York: Rizzoli, 1981.

Libeskind, Daniel, *Chamberworks*, London: Architectural Association, 1983.

Libeskind, Daniel, *Theatrum Mundi*, London: Architectural Association, 1985.

Libeskind, Daniel, *Line of Fire*, Milan: Electa, 1988.

Libeskind, Daniel, *Countersign*, London: Academy Editions, and New York: Rizzoli Editions, 1992.

Libeskind, Daniel, *Kein Ort an seiner Stelle*, Dresden: Verlag der Künste, 1995.

Libeskind, Daniel, *Fishing from the Pavement*, Rotterdam: NAI Uitgevers/Publishers, 1997.

Libeskind, Daniel, *The Space of Encounter*, New York: Universe Press, 2001.

Libeskind, Daniel and Balmond, Cecil, *Unfolding*, Rotterdam: NAI Uitgevers/Publishers, 1997.

Libeskind, Daniel, *et al.*, *The Jewish Museum Berlin*, Berlin: Verlag der Kunst, 1999.

Müller, Alois Martin (ed.), *Radix: Matrix: Works and Writings of Daniel Libeskind*, Munich: Prestel Verlag, 1994.

Müller, Alois Martin (ed.), *Radix: Matrix: Works and Writings of Daniel Libeskind*, Munich: Prestel Verlag, 1997.

Rodiek, Thorsten, *Museum ohne Ausgang: Das Felix-Nussbaum-Haus des Kulturgeschichtlichen Museums Osnabrück – Daniel Libeskind*, Tübingen: Wasmuth Verlag, 1998.

Sacchi, Livio, *Universale di architettura. Daniel Libeskind, Museo ebraico, Berlino*, Turino: Testo & Immagine, 1998.

Schneider, Bernhard, *The Jewish Museum Berlin, Daniel Libeskind*, Munich: Prestel Verlag, 1999.

4 Virtual space

CHAR DAVIES

> *What I am trying to translate to you is more mysterious; it is entwined in the very roots of being, in the impalpable source of sensations.*
>
> J. Gasquet, *Cézanne*, quoted by Merleau-Ponty, 'Eye and mind'

I have been working in 'virtual space' for nearly ten years, and during that time have produced two major works, the virtual environments *Osmose* (1995) and *Ephémère* (1998).[1] Integrating full-body immersion, interactive 3-D digital imagery and sound, and navigation via a breathing interface, these works embody a radically alternative approach to immersive virtual space, or what is commonly known as 'virtual reality' or 'VR'. Rather than approaching the medium as a means of escape into some disembodied techno-utopian fantasy, I see it as a means of return, i.e. of facilitating a temporary release from our habitual perceptions and culturally biased assumptions about being in the world, to enable us, however momentarily, to perceive ourselves and the world around us *freshly*.

It should be noted that when I say virtual space, I am referring to *immersive* virtual space, i.e. a computer-generated artificial environment that one can seemingly, with the aid of various devices, go inside. I think of virtual space as a spatiotemporal 'arena' wherein mental models or abstract constructs of the world can be given virtual embodiment (visual and aural) in three dimensions

---

[1] *Osmose* (1995) and *Ephémère* (1998) were constructed with the dedicated participation of the following individuals: John Harrison, custom programming; Georges Mauro, graphics; Dorota Blaszczak, 3-D sonic architecture; and Rick Bidlack, sound composition.

Space: In Science, Art and Society, edited by F. Penz, G. Radick and R. Howell.
Published by Cambridge University Press. © Darwin College 2004.

FIGURE 4.1 Forest Stream, *Ephémère*, 1998. Digital image captured in real-time
through head-mounted display during live immersive journey/performance.

and be animated through time. Most significantly, these can then be kinesthetically explored by others through full-body immersion and real-time interaction,
even while such constructs retain their immateriality. Immersive virtual space
is thus a philosophical *and* a participatory medium, a unique convergence in
which the immaterial is confused with the bodily-felt, and the imaginary with
the strangely real. This paradox is its most singular power. The firsthand experience of being bodily immersed in its all-encompassing spatiality is key: when
combined with its capacity for abstraction, temporality and interaction, and
when approached through an embodying interface, immersive virtual space
becomes a very potent medium indeed.

Between 1995 and 2001, more than 20 000 people were individually
immersed in the virtual environments *Osmose* and *Ephémère*. A common
response to the experience is one of astonishment: many 'immersants' have
described their experience in euphoric terms while others have inexplicably

wept. As one participant wrote six months afterwards: '[This experience] heightened an awareness of my body as a site of consciousness and of the experience and sensation of consciousness occupying space. It's the most evocative exploration of the perception of consciousness that I have experienced since I can't remember when.'

Such responses suggest that immersive virtual space, when approached in an unconventional way, can indeed provide a means of perceiving freshly. The medium's paradoxical qualities may effectively be used to redirect attention from our usual distractions and assumptions to the sensations of our own condition as briefly embodied sentient beings immersed in the flow of life through space and time.

Virtual space and King Logos

Many centuries after Copernicus and Galileo's dismantling of the terra-centric universe, we still refer to the sun rising and setting on the horizon as if the earth were flat; similarly, many decades after Einstein's relativity theory, in everyday life we continue to conceptualise the world around us in terms of the old Newtonian/Cartesian paradigm, i.e. as an aggregate of solid separate objects in empty space. As Roger Jones wrote in *Physics as Metaphor* (1982):

> The modern notion of space is a compound metaphor that embodies all our concepts and experiences of separation, distinction, articulation, isolation, delimitation, division, differentiation and identity. The laws of perspective and of geometry for us are a codified summary of our normal experience of alienation, unique identity, and un-relatedness. It has all been abstracted, externalized, and synthesized into the cold, empty void we call space. This metaphor of space is our modern mechanism for avoiding the experience of oneness, of the chaos, of the ultimate state of unity to which the mystic seers and philosophers of all ages have referred.

Conventional ways of thinking about and producing immersive virtual space faithfully mirror this metaphor. 3-D computer graphic techniques, as commonly used in VR environments, tend to rely on 3-D Euclidian geometric models, Renaissance perspective and the xyz coordinates of Cartesian space, all applied in a never-ending quest for visual realism. The resulting aesthetic/sensibility (what I call the 'hard-edged-objects-in-empty-space' syndrome) reflects a dualist, objectifying interpretation of the world. When these techniques are combined with what have already become conventional methods of user inter-action (such as hand-held joysticks, pointers, gloves, etc.) the effect – regardless of content – reinforces a particular way of being in the world in terms of mastery, domination and control.

It is important to understand that virtual space is not neutral. The ori-gins of the technology associated with it lie deep within the military and Western-scientific-industrial-patriarchal complex. It should not be surprising then if the medium not only reflects these values but, by default, reinforces what Henri Lefebvre, in *The Production of Space* (1991), calls the reign of King Logos:

> King Logos is guarded on one hand by the Eye – the eye of God, of the Father, of the Master or Boss, which answers to the primacy of the visual realm with its images and its graphic dimension, and on the other hand by the phallic (military and the heroic) principle, which belongs, as one of its chief properties, to abstract space.

In its most prevalent form, virtual reality can thus be considered a 'literal re-enactment of Cartesian ontology', as Richard Coyne wrote in 'Heidegger and virtual reality: the implications of Heidegger's thinking for computer rep-resentations' (1994). In conventional VR, the participating human subject is represented as an omnipotent, disembodied and isolated view-point, manoeu-vring in empty space (and often, at least in terms of increasingly immersive computer games, looking for something to kill . . .). Numerous other writers have analysed the cultural bias inherent in the medium. For example, Ziauddin Sardar, in 'Cyberspace as the darker side of the West' (1996), has called virtual reality a product of the collective unconscious of Western culture, suggesting it issues from 'a techno-utopian ideology ripe with subconscious perceptions and prejudices', in which 'liberation from the body is sought by dissolving into the machine'.

VR's tendency toward disembodiment should not be surprising either. As a realm ruled by mind, virtual reality – as conventionally constructed – is the epitome of Cartesian desire, in that it enables the construction of artificial worlds where there is the illusion of total control, where ageing mortal flesh is absent, and where, to paraphrase Laurie Anderson, there is no 'dirt'. I believe such a desire to escape the confines of the body and the physical world is symptomatic of an almost pathological denial of our embodied embeddedness in the living world. It is tempting to suggest that belief in artificial intelligence and silicon as a means of delivery into immortal omnipotence in some other Eden is but a testosterone-induced dream.

In the virtual environments *Osmose* and *Ephémère*, I have proposed an *alternative* approach to virtual space, intended to resist the cultural trajectory described above. With this intent, we have developed strategies such as an embodying user-interface which grounds the immersive experience in the participant's own breathing and balance. We have also employed semi-transparency in the visuals so as to create a perceptual ambiguity which might serve to dismantle the Western 'mis-perception' of the world.

My desire to accomplish this task, to propose an alternative, is rooted in my own particular experience of being in the world. Most importantly, this desire, and the strategies developed, have evolved through many years of artistic research into my own perception of light and space. I should also add that a decade spent within the software industry (1987–97) as a founding director and head of visual research at a world-leading software development company (Softimage, whose software tools were used in Hollywood movies such as *Jurassic Park* and *The Matrix*) made me acutely aware of the technology's bias toward reinforcing a traditional Western worldview. This awareness further fuelled my desire to push the technology and prove that it could indeed be used to express a different sensibility.

An alternative sensibility: a spatiality without things

> How would the painter or poet express anything other than [her] encounter with the world?
>
> Merleau-Ponty, *Signs*

I came to the medium of immersive virtual space as a painter, seeking a more effective means of communicating my sensibility of the world. My lifelong artistic project (now stretching over twenty-five years) has been to re-present the world as I have intuitively sensed it to be – *behind the veil of appearances* – as immaterial, interrelated and dynamic flux. Within this all-enveloping flux and flow, habitually perceived distinctions between things dissolve, and boundaries between interior self and exterior world become permeable and intermingled.

FIGURE 4.2 *Blue World Space.* Oil/acrylic on canvas, 1985. Char Davies.

This quest, to understand my intuition further and to articulate it effectively to others, is the driving force behind my work: *Osmose* and *Ephémère* are the most recent fruits of this endeavour.

Many of the strategies and aesthetic principles I have employed in my work are grounded in my own physiological experience of vision. My eyes are extremely myopic (at 17 'diopters', in layman's terms they require a thickness of 17 corrective lenses to see the world in focus with close to the same 20/20 'Mc-vision' acuity as everyone else). When 'uncorrected' through prescription lenses, I encounter a radically different spatiality, in which normally perceived boundaries between objects and surrounding space are dissolved in light. Here, all semblance of hard edges, all sense of solid-surfaced separate objects, and all distinctions between things, including figure and ground, near and far – the usual perceptual cues by which we objectify the world – simply disappear, dissolved into an ambiguous enveloping spatiality of soft, semi-transparent, intermingling volumes of varying hues and luminosities. This unusual spatial sensibility bears a striking resemblance to Merleau-Ponty's description of night in *The Phenomenology of Perception* (1962):

> When, for example, the world of clear and articulate objects is abolished, our
> perceptual being, cut off from its world, evolves a spatiality without things.
> This is what happens in the night. Night is not an object before me; it enwraps
> me and infiltrates through my senses, stifling my recollections, and almost
> destroying my personal identity. I am no longer withdrawn into my perceptual
> look-out from which I watch the outlines of objects moving by at a
> distance . . . it is pure depth without foreground or background, without
> surfaces and without any distance separating it from me.

In my own experience, such withdrawal of visual acuity – which so dom-
inates our habitual perception of space – allows another way of 'sensing' to
come forward, just as Merleau-Ponty suggests. This is essentially a spatiality without 'things', in which the threshold between interior self and exterior
world becomes porous, and the separation between 'out there' and 'in here' is
transcended. Whereas visual acuity tends to keep attention focused on what
lies in front or ahead (i.e. the future), when it dissolves into a non-focused
blur one becomes aware of space as *all around*, bodily enveloping as if one
were immersed in the sensuous liquidity of the sea. Merleau-Ponty, in 'Eye and
Mind' (1964), also describes such space:

> no longer . . . a network of relations between objects such as would be seen by a
> witness to my vision or by a geometer looking over it and reconstructing it from
> the outside. It is, rather, a space reckoned starting from me as the zero point or
> degree zero of its spatiality. I do not see it according to its exterior envelope; I
> live it from the inside; I am immersed in it. After all, the world is all around me,
> not in front of me.

When visual acuity is decreased, one also becomes more aware of sound; and
sound, as an all-encompassing flux which penetrates the boundary of the skin,
further erodes the distinctions between inside and outside. As the Australian
sound theoretician Frances Dyson said at a conference in 1994, 'metaphysically,
sound has an ontology that challenges the solid world'. Sound, like soft vision,
also returns us to what I have come to call the 'presence of the present'. In this
perceptual state, rather than being mentally focused on the future and thus
inattentive, even absent, to the present, one becomes acutely aware of one's own
embodied presence inhabiting space, in relation to a myriad of other presences
as well.

Many of the key characteristics associated with my work, such as full-
body immersion in an all-surrounding visual and auditory space, and the

FIGURE 4.3 *Logger & Tree*, 1981, oil on canvas.

semi-transparent, dematerialising quality of the visuals, are thus grounded in my own experience of vision. While I could defer to various theoretical analyses of spatial perception, it is really through years of artistic investigation into my own bodily experience of space, through painting, that I have gained such insights.

From painting to immersive virtual space

While I began my career as a realist painter, taking great satisfaction in depicting the hard-edged boundaries between things, a chance turning of artistic attention to my own perception of space led me to acknowledge, in 1981, this alternative sensibility. Subsequently, I began making studies from life without wearing corrective lenses. Over the years, this led to the development of the visual aesthetic of semi-transparency and semi-abstract/semi-representation used in *Osmose* and *Ephémère*. Figures 4.3–4.6 show several images I produced during that time, beginning with an example of one of my last realist works. In the still-lifes, I was exploring the dissolution of form through light, and the

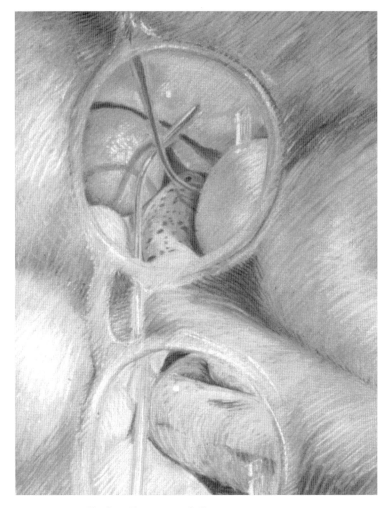

FIGURE 4.4 *Still Life & Glasses*, 1981, chalk on paper.

ensuing erosion of perceivable boundaries between things. These studies cul-minated in paintings of glass jars on mirrors in which I was essentially painting the flow of light in volumous space.

This research eventually led to another, more abstract, body of work created between 1985 and 1987 and exhibited as *Espaces Entrelacés (Interlaced Space)*. I considered these paintings, some included here, as landscapes even though they were not created 'from life'. In these images, I was attempting to convey

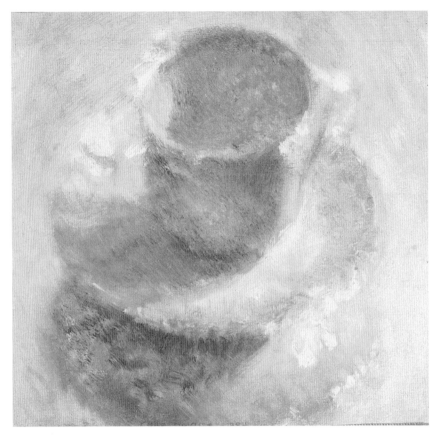

FIGURE 4.5 *White Cup*, 1984, detail, oil on board.

the experiential intermingling of the exterior flowing world and the interior perceiving embodied self, as well as to communicate the subjective sensation of being enveloped in all-encompassing space.

Eventually, however, the two-dimensionality of the painterly picture plane posed an insurmountable limitation, because I could not effectively articulate the sense of being all-enveloped, nor could I convey flux and flow. In the mid 1980s, I saw an example of early 3-D computer animation, consisting of phosphorescent green vector graphics against black space. In that short clip, I recognised the potential of the medium for my own purposes, and by the end of 1987 had become a founding director of the software company Softimage. I was interested not in the computer technology itself, but in the possibility of using it

FIGURE 4.6 *Glass Jars & Mirror*, 1985, oil on canvas.

to create on the 'other side' of the picture plane. Within a few years, I produced a series of 3-D digital images, collectively titled *Interior Bodies* (1990–3), which explored the metaphorical co-equivalence of nature and body. In these images, I adapted my previous painting techniques and used the software's lighting and transparency effects to circumvent the hard-edged polygonal models so characteristic of the technology and to create instead the soft spatial ambiguity I desired.

While these works were created with 3-D software, they were reproduced as 2-D and static images, thus defeating my original intent. Intuiting that the immersive space of virtual reality might offer a more effective means of articulating my sensibility – and provide a way of enabling my audience to 'cross over' the 2-D picture plane with me – in 1993 I began to conceptualise an immersive virtual environment and put together a team. This work became *Osmose*.

FIGURE 4.7 *Blue world space*, 1985, oil on canvas.

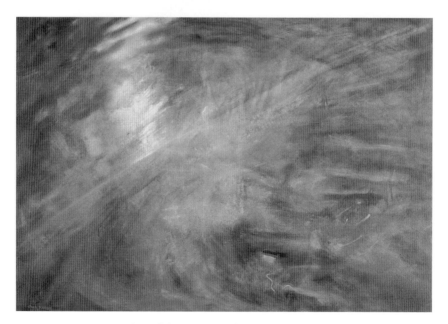

FIGURE 4.8 *Beyond the cave*, 1987, oil on canvas.

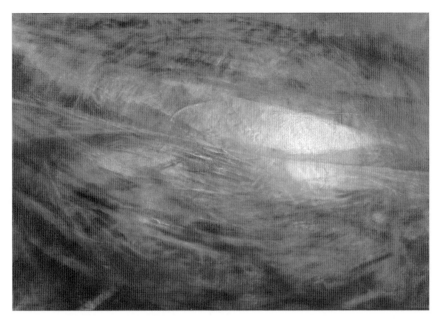

FIGURE 4.9 *Lake*, 1987, oil on canvas.

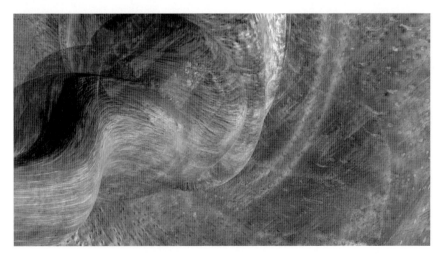

FIGURE 4.10 *Root*, 1991, digital image made with 3-D software, duratrans colour transparency in light box – approx. 4 ft × 5 ft.

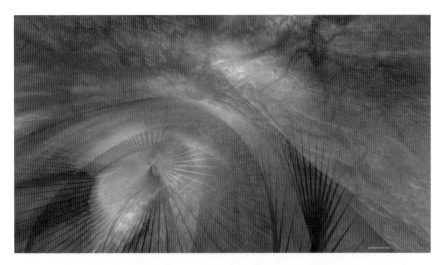

FIGURE 4.11 *Seed*, 1991, digital image made with 3-D software, duratrans colour transparency in light box – approx. 4 ft × 5 ft.

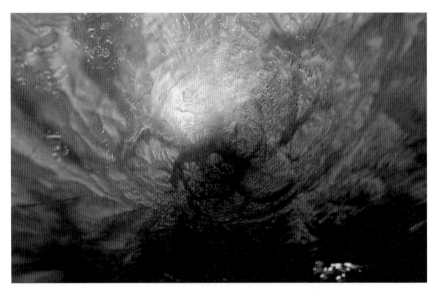

FIGURE 4.12 *Yearning*, 1993, digital image made with 3-D software, duratrans colour transparency in light box – approx. 4 ft × 5 ft. See also colour insert, Plate 1.

The virtual environments *Osmose* and *Ephémère*

In the following pages, I will describe *Osmose* and *Ephémère*. But first, a caveat: these are bodily experiential works. As such they cannot easily be described in words or documented by the 2-D images that accompany this text: rather they are a mode of access to an ephemeral yet embodied experience of self and space, or rather *self in place*. To understand, one must really experience them firsthand: just as the sensations of becoming wet and buoyant can only be known by the swimmer, so these works can only truly be understood through subjective bodily participation.

To access their virtual spatiotemporal realms (at least in the current era, for the following methods will evolve as technology develops) one must dress in specific gear, as divers do. This includes donning an interface vest and a stereoscopic viewing helmet (known in the field as a head-mounted display or HMD). Inside the helmet are two small LCD screens which together create a stereoscopic effect, as well as stereo headphones. The vest and HMD are linked through various cables to a computer and digital sound synthesisers/processors. As one looks around (including behind one's back and below one's feet), the computer calculates one's point of view and relative changing spatial position within the virtual realm via motion-tracking sensors in the interface vest and helmet, and, in response, generates the appropriate visual elements and aural effects in real-time, i.e. *on the fly*.

To navigate within *Osmose* and *Ephémère*, all one needs to do is breathe – breathing in to rise, out to fall – and shift one's centre of balance and lean in order to change direction. More technically speaking, we accomplished this by placing motion sensors on the participant's vest, to: (a) track the expansion and contraction of the chest as the immersant engages in breathing; and (b) track the relative tilt of the spine as the participant leans one way or another.

This strategy, of having the immersive experience dependent on the intuitive visceral processes of breath and balance, was intended to counter conventional ways of navigating and interacting in virtual space. (Such techniques, by relying on hand-based devices such as joysticks, pointers or data gloves, tend to reinforce an instrumental, dominating stance toward the world.) Our approach was intended to counter the medium's bias with a vision of the medium as a channel for 'communion' rather than control.

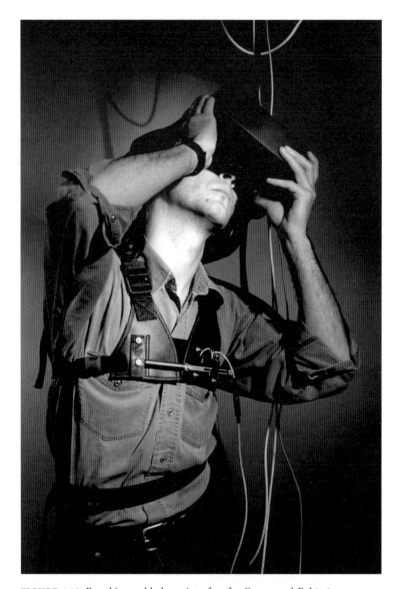

FIGURE 4.13 Breathing and balance interface for *Osmose* and *Ephémère*.

Our use of the participant's own breathing and balance for interface has many implications. As Drew Leder wrote in *The Absent Body* (1990):

> Breath is a potent tool of overcoming dualism. Physiologically, respiration stands at the very threshold of the ecstatic and visceral, the voluntary and the involuntary . . . inside and outside, self and Other are relativized, porous, each time one takes a breath. The air is constantly transgressing boundaries, sustaining life through inter-connection. One may have spent years studying the mystics on the unreality of dualism and this remain an abstract idea. But in following breath, one begins to embody this truth.

> Balance is a question of centering. When we are properly centered, our experience of Being is in equilibrium. Being well-centered, we can encounter other beings in a more open, receptive way. Finding our center is a necessary step in the development of our ontological capacity to open ourselves to the larger measure of being and to encounter other beings with a presence that is deeply responsive. Coming home to our true center of being, we can begin to relax our egological defences, and begin to experience things outside the subject/object polarization.

In *Osmose* and *Ephémère*, the experience of breathing in to rise and out to fall facilitates a convincing sensation of 'floating', as if the participant's body were gravity free. This unusual sensation is intimately known by scuba divers, who use breath and balance to control body buoyancy subtly and to manoeuvre in oceanic space. In *Osmose* and *Ephémère*, the sensation of floating tends to evoke euphoric feelings of disembodiment and immateriality, which we intentionally amplify through our enabling the participants to see through and virtually float through everything around them. At the same time, however, we deliberately confound these sensations by paradoxically grounding the experience in the participant's own body, i.e. in his or her own breath and centre of balance. In this way, *Osmose* and *Ephémère* seek to reaffirm the presence, often overlooked or denied in conventional VR, of the subjectively inhabited body in immersive virtual space.

As a means of subverting the conventional VR aesthetic of hard-edged-objects-in-empty-space, we use semi-transparency and translucency in the visuals, an approach developed long before in my painting. Thus, when an 'immersant' is within *Osmose* or *Ephémère*, everything he or she sees is semi-transparent. Just as in my own un-'corrected' vision, there are no sharp distinctions between solid bounded objects in foreground and background, and no empty space, but instead, ever-changing abstractions of semi-transparent forms. The effect for the immersant is of floating within a world which is neither wholly representational (i.e. recognisable) nor wholly abstract, but hovering in between. As the participant moves within the virtual space, the ever-changing spatial relationships between the various semi-transparent forms (one can see through more than twenty layers simultaneously, a major technical challenge at the time) create a constantly changing variability of the perceptual field. This generates semiotic and sensory fluctuations or what I simply call 'perceptual/conceptual buzz'. Based on a painterly strategy of maintaining a 'razor's edge' between representation and abstraction, whereby multiple associations or interpretations are deliberately evoked (rather than a single meaning being literally illustrated), our intent was to heighten ambiguity in order to refocus the participant's attention on their own act of perceiving, or rather of being.

In *Osmose* and *Ephémère*, the immersive experience is also significantly affected by our use of sound. As one journeys throughout the spatial realms, one is immersed in constantly changing sound coming from all directions. (In *Osmose* the sounds are derived from a male and female human voice uttering phonetics, and in *Ephémère* from a viola, digitally altered to create a vast range of aural effects.) The sounds have been 'localised' in three dimensions and have been designed to transform, like the visuals, on the fly, in real time, in response to the immersant's ever-shifting position, speed, direction of gaze, and various other events. In both works, the sounds have been composed to oscillate between melodic form and mimetic effect in a state somewhere between structure and chaos.

In terms of content, both *Osmose* and *Ephémère* are based on nature and landscape as metaphor. As such, their realms are populated with trees, roots, rocks, streams, etc., all iconic elements which have reoccurred in my work for twenty-five years. It is outside the scope of this chapter to discuss the reasons for and implications of re-presenting the natural world in virtual environments. However I do want to emphasise that my intention has been to use the medium's unique qualities to present nature beyond the veil of surface appearances, while grounding such perception in the subjectively lived body. In this context, my work could be interpreted as an ongoing attempt to articulate a vision of nature perhaps closer to how Heidegger (in *Heraklit*) described the Greek's 'physis' – as 'outside of all specific connotations of mountains, sea or animals, the pure blooming in the power of which all that appears and thus "is"'.

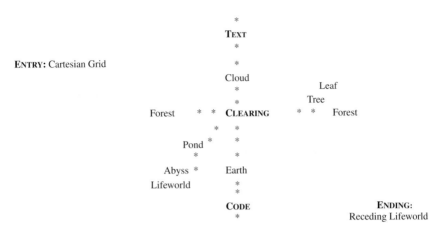

```
                                *
                              TEXT
                                *

ENTRY: Cartesian Grid           *
                              Cloud                 Leaf
                                *                   Tree
Forest      *    *    CLEARING     *    *    Forest
                    *    *
            Pond  *      *
                    *    *
         Abyss  *     Earth
         Lifeworld       *
                         *
                      CODE                ENDING:
                         *            Receding Lifeworld
```

FIGURE 4.14 Spatial structure of *Osmose*.

Before going on to describe immersive journeys through *Osmose* and *Ephémère*, I want to emphasise that they are spaces, or rather *places*, for perceptual play. They do not contain a predetermined linear narrative. In these works, each participant's experience is unique, unrepeatable, dependent on one's own behaviour, on one's whim or will. In the following pages, I am going to describe *Osmose* and *Ephémère* as if I were leading you on a journey; but it is important to remember that I could take you on many different trajectories through these two works, because they are places, virtual landscapes, through which the participant may roam, engaged in solitary reverie.

A journey through the virtual realm of *Osmose*

Osmose: derived from 'osmose' (Fr.), 'osmosis' (Eng.) from 'osmos' (Grk) 'to push'; a biological process involving passage from one side of a semi-permeable cellular membrane to another. *Osmose as metaphor*: a transcendence of difference through mutual absorption; a dissolution of boundaries between inner and outer; an inter-mingling of self and world.

Osmose consists of nearly a dozen realms, of forest, pond, subterranean earth and so on, all situated around a central clearing. The spatial structure of the work has a strong vertical axis (rather than the conventional horizontal plane of most VR works) – amplified by the use of breath to rise buoyantly or descend. Vertically, there is a kind of spatial recycling, whereby if the participant ascends

FIGURE 4.15 Tree pond, *Osmose*, 1995, digital frame captured in real-time through head-mounted display during live performance of immersive virtual environment. See also colour insert, Plate 2.

to the very heights of the space she will be returned to its depths, and vice versa.

When an immersant first 'enters' *Osmose*, he or she will find him/herself in the midst of a 3-D grid extending infinitely in empty black space. This grid (a reference to the Cartesian xyz coordinate system) functions as an orientation site for becoming familiar with the breath and balance interface. The immersant will soon realise she is buoyantly 'floating' as if gravity free, rising and falling according to the rhythms of her own breath, and that she can hover in 'mid-air' or glide, as well as change direction by shifting her centre of balance.

A few moments after her entry, the grid soon fades, leaving the immersant in the middle of a clearing. Gazing all around, she sees, or perhaps first hears, what appears to be a sienna-hued oak tree, near a small pond into which is flowing a stream of light particles, and all around, a circumference of dark

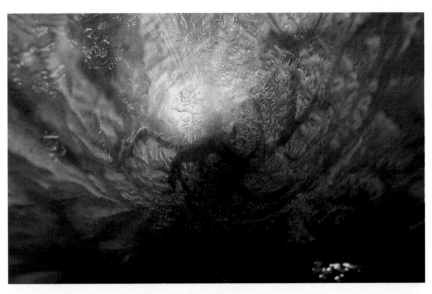

PLATE 1 *Yearning*, 1993, digital image made with 3-D software, duratrans colour transparency in light box – approx. 4 ft × 5 ft.

PLATE 2 Tree pond, *Osmose*, 1995, digital frame captured in real-time through head-mounted display during live performance of immersive virtual environment. Char Davies.

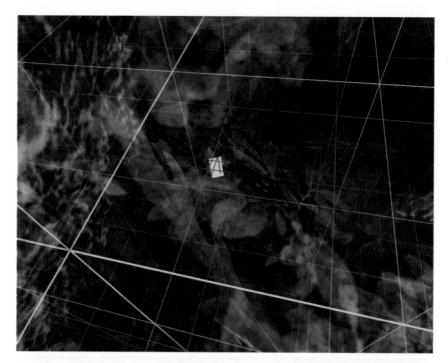

PLATE 3 Forest grid, *Osmose*, 1995, digital frame captured in real-time through head-mounted display during live performance of immersive virtual environment.

PLATE 4 Winter swamp, *Ephémère*, 1998, digital image captured in real-time through head-mounted display during live immersive journey/performance.

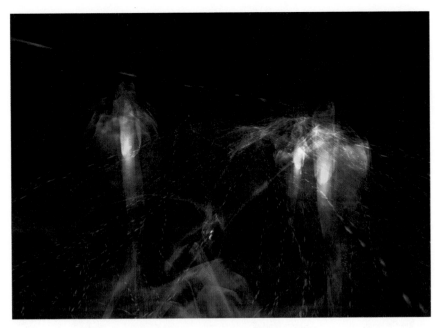

PLATE 5 Seeds, *Ephémère*, 1998, digital image captured in real-time through head-mounted display during live immersive journey/performance.

PLATE 6 An immersant in *Osmose*, seen through the shadow silhouette screen.

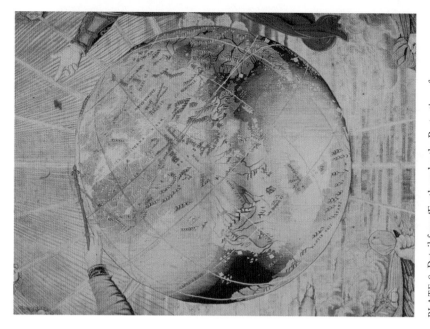

PLATE 8 Detail from 'Earth under the Protection of Jupiter and Juno' (Plate 7).

PLATE 7 Third tapestry of the 'Spheres' series of three: 'Earth under the Protection of Jupiter and Juno', attributed to Bernard van Orley. Brussels, c. 1520–30, gold, silver, silk and wool, 11ft 2ins × 11ft 4ins, © Patrimonio Nacional, Madrid.

forest. The clearing cycles through day and night, its ebb and flow of light and dark accompanied by subtle visual and aural changes. These include, at night, flitting firefly-like lights and a nocturnal melody repeating over and over (created, like every other sound in *Osmose*, from phonetic utterances digitally altered).

The immersant will realise she has entered a non-Cartesian place, very unlike the 'real world': here, everything is dematerialised and semi-transparent – there are no solid surfaces, no hard edges, no separate objects in empty space. Instead, the immersant can see through everything – through the body of the tree, the ground, the roots below.

She may choose to drift into the clearing's tree, rising with its streaming particles. Or she can float into its branches, only to find herself passing through a previously invisible leafy canopy and into the interior of a leaf, consisting of brightly blazing lights streaming through green space, accompanied by high-pitched sound. From within the clearing, if the immersant breathes shallowly and leans forward, she can also glide toward the encircling forest. As she nears its edge, the clearing will fade and the forest realm will begin to appear all around her. For several moments, she may find herself in a non-Cartesian spatial intermingling of clearing and forest, in which she is paradoxically enveloped by both realms at once (with skill, it is possible to remain within this strange liminal zone, although moving forward or back will summon in one realm and cause the other to fade).

Once within the forest, the immersant is surrounded by a thick mass of large semi-transparent leaves (created by digitally scanning real leaves – the only use of the 'real' in the entire work, all else is digitally constructed). As she floats, these leaves constantly re-form themselves around her, creating an endlessly recurring space. In the *Osmose* forest, heading in a straight line will only cause it to recur forever (and moving too quickly will summon in the Cartesian grid). It is possible to exit the forest by following a stream of flowing luminous particles back to the clearing's pond, or, alternatively, by remaining still and hovering in one place: this causes the forest to fade and the clearing to reappear.

From the clearing, the immersant can also approach the pond (perhaps guided to its location by its emitting of frog-like sounds) and hover above its transparent surface. If sufficiently deft in use of breath and balance, she can

FIGURE 4.16 Tree, *Osmose*, 1995, digital image captured in real-time through head-mounted display during live immersive journey/performance.

descend through its lowest depths into an oceanic abyss. This seemingly vast space is populated only by dimly visible streaming fishlike entities far below, as well as echoing calls all around. Eventually the immersant will hear a distinctive tinkling sound behind her. To leave the abyss, she must head toward that sound into the reference-less big blue. In response, a translucent pod-like entity, the 'lifeworld', will appear. As she moves toward it, it also tumble towards her until she is engulfed within it: she may realise that the lifeworld was the clearing

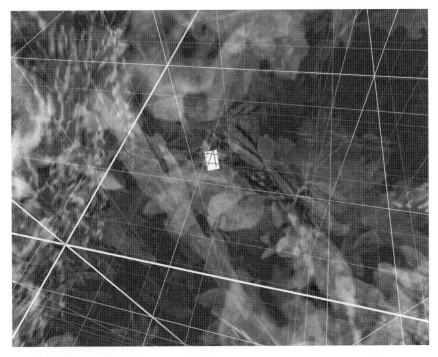

FIGURE 4.17 Forest grid, *Osmose*, 1995, digital image captured in real-time through head-mounted display during live performance of immersive virtual environment. See also colour insert, Plate 3.

seen from without as a miniaturised globe, and that now she is back inside it, hovering above the clearing's tree.

The immersant can also look down directly into the subterranean earth, because the ground too is semi-transparent. If she exhales deeply, she can descend vertically into this realm, aurally resonant and populated by the ghostly forms of semi-transparent, semi-abstracted roots and boulders and luminous particle streams. The immersant can glide through this realm, floating among and through its various elements, and she can depart simply by taking a breath to ascend back to the clearing.

If the immersant so desires, she can descend even deeper to another realm below the earth, of software code. Represented here are thousands of lines of phosphorescent green text in black space, the actual code John Harrison programmed to construct the work. It is possible to float among these scrolling walls of code, including the constantly changing measurements of the

FIGURE 4.18 Subterranean earth, *Osmose*, 1995, digital image captured in real-time through head-mounted display during live immersive journey/performance.

immersant's own breathing. The code realm was intended to function as the conceptual substrate of *Osmose*, drawing attention to the computer-generated artificiality of the experience.

The immersant can also ascend from the clearing through a thick ceiling of whistling cloud into a realm of pale fog within which float scrolling passages of written text. These are excerpts of writing on perception, space, nature, the

FIGURE 4.19 Code world, *Osmose*, 1995, digital image captured in real-time through head-mounted display during live immersive journey/performance.

body, and technology by poets and philosophers such as Rainer Maria Rilke, Gaston Bachelard, and Merleau-Ponty, whose ideas have accompanied and encouraged me in the past decade. It is possible to float in the grey fog among these scrolling texts and listen to their accompanying aural effects, which sound somewhat like choral voices even though, like all the sounds in *Osmose*, they are derived from a single male and female voice uttering phonetics. The text realm functions as the conceptual superstrate of *Osmose*, and together with the code, its twin in the underworld, provides a conceptual framework around the entire creation.

After a certain amount of time has elapsed (we limit the immersive session to 15 minutes, even though one could stay in *Osmose* indefinitely) the journey is gently brought to a close. The ending is signalled by a reoccurrence of the clearing's nocturnal melody, and a gentle reappearing of the lifeworld. Immediately however, the lifeworld begins to recede, irretrievably, until it has shrunk to a tiny spec and then disappears altogether, leaving the immersant floating alone in empty dark space.

| ENTRY: WINTER SWAMP (PROLOGUE) | | SPRING | > | SUMMER | > | AUTUMN > | | ENDING |

Entry: WINTER Swamp (Prologue)   SPRING   >   SUMMER   >   AUTUMN >   ENDING
*
Forest Landscape: River,   Boulders, Foliage leafing   > Falling trees & leaves > embers
*   & ash
Subterranean Earth: Stream,   Boulders, Seeds germinating > Decay   > embers
*   & ash
Interior Body: Arteries, Organs,   Eggs falling   > Bones (aging)   > embers
*

FIGURE 4.20 Spatial and temporal structure of *Ephémère*.

A journey through the virtual realm of *Ephémère*

In comparison to *Osmose* – which I think of somewhat as a perceptually mesmerising stage set (whereby motion is primarily derived from the immersant's buoyant passage and resultant ever-changing spatial relationships throughout) – I think of *Ephémère* more like a virtual 'opera'. For *Ephémère* is a temporal space, whereby every element, every form, has been choreographed (within a range of randomness) to engage in constant transformation, in an unceasing ebb and flow, wax and wane of visibility and audibility. As the name *Ephémère* suggests, this work was intended as an evocation of ephemerality: of the fleeting quality of our own lifespans as mortal beings, embodied among an unfathomable myriad of other beings, all engaged in coming-into-being, lingering and passing-away in the flow of life through space-time.

Ephémère is structured spatially on a vertical axis, with three horizontal levels: forest landscape, subterranean earth, and interior body. Accordingly, the work's iconographic repertoire has been expanded beyond *Osmose*'s trees, boulders, pond and stream to include flesh, bloodstreams and bone. However, even as the immersant roams among all three realms in *Ephémère*, no realm remains the same. The forest landscape changes continually, passing through cycles of day and night, and transforming through the seasons. Deep within the earth, huge boulders transform into pulsing body organs, and, within the body, eggs appear, while ageing organs give way to bone. Throughout the duration of the experience, the various elements of rocks, roots, seeds etc., come into being, linger and pass away: the timing of their appearings and disappearings is dependent not only on the temporal progression of the work, but on the immersant's vertical position, proximity, slowness of movement, and steadiness/duration of gaze. (For example, when gazed upon, boulders summon phantom landscapes, and seeds germinate, inviting entry into the luminous interior of their blooming.)

Here too, there is a stream or river, but unlike the stream in *Osmose*, the one in *Ephémère* has a gravitational pull and provides an alternative means of navigating within the space. If the immersant floats too near, the river will suck her into its force and carry her along. It is actually possible to experience much of *Ephémère* by passively submerging oneself in the swiftly flowing and noisy stream / underground river / artery: for if the immersant remains within it for a sufficient amount of time, it will respond by randomly transforming the surrounding spatial realm into one of the other realms of *Ephémère*.

When a participant first enters *Ephémère* (wearing the same gear as worn in *Osmose*), she will find herself floating among star-like points of light in dark space. As she becomes accustomed to breathing in to rise and out to fall, the particles of light begin to fall like snow, and the darkness gives way to fog, filled with aural effects of wind and rattling wood. Far below are dark slender forms of trees. Upon exhaling, the immersant may drift down among them, their branchless shapes suggesting they are but relics, no longer living. There are also semi-transparent boulders. By now she will have heard, and then perhaps turned to see, a darker horizontal ribbon with bright particles flowing within it, intended to suggest a winter stream. The immersant, still enveloped in ebbing grey light and falling snow, is in the winter swamp, the 'prologue' of *Ephémère*.

In the winter swamp, everything is more abstract than in the *Osmose* clearing, as well as being in black and white rather than in colour. Here (unlike *Osmose*'s spatially anchoring tree to which immersants can return for reassurance and to regain their bearings) no place remains the same: every element is engaged in a constant process of coming into being, lingering, and passing away, both visibly and aurally. It is almost as if the immersant has floated down into a ghostly realm engaged in its own processes of becoming and un-becoming. If the immersant chooses to float toward a tree and passes through its slender trunk, in response she will hear a cry like distant ripping wood as an almost transparent white tree (the *Osmose* tree making a cameo) slowly falls and crashes. If she passes through other trees, this ghost tree will fall again, and again.

If the immersant slowly approaches one of the boulders (which like the trees are engaged in their own appearings and disappearings) and steadily gazes at

FIGURE 4.21 Winter swamp, *Ephémère*, 1998, digital image captured in real-time through head-mounted display during live immersive journey/performance. See also colour insert, Plate 4.

it from a tactful distance, in response the boulder will summon in a phantom landscape. This landscape, with its encircling horizon of Arctic-like bergs of ice and snow, and its accompanying sounds, will briefly appear all around her, then fade, leaving her in the swamp. Meanwhile, the ambient light is changing as dusk ebbs into night. The entire swamp has been 'choreographed' as a prologue lasting several minutes, and the immersant is unable to move elsewhere until night eases into dawn.

After a certain amount of time has elapsed, the swamp transforms itself into the forest landscape realm, composed not only of trees and boulders and a river, but of abstract intermingling layers of various greens and vertical flows of luminous particles, all engaged in various rates of appearing and disappearing. As in the winter swamp, the forest's boulders are also 'sensitive' to the immersant's behaviour, and, if approached in a certain way, will summon in another

FIGURE 4.22 Autumn forest landscape, *Ephémère*, 1998, digital image captured in real-time through head-mounted display during live immersive journey/ performance.

phantom landscape which fades almost as soon as it appears. The forest itself is transforming constantly, irrepressibly engaged in a seasonal progression from the snowy pale of winter through spring and summer to the climactic decay of autumn. The immersant can choose to spend her entire journey in the forest, but if she leaves it perhaps to return later, it will continue to progress over time.

As in *Osmose*, if the immersant exhales deeply, she will descend vertically through the semi-transparent ground, past the spreading roots of trees into a vast subterranean earth. Here she can hear deep rumbling sounds of shifting rock, as seemingly gigantic boulders drift, and slowly appear and disappear. The comings and goings of these rocks have all been choreographed, some in groups, some individually, at different depths. They are richly hued in ochres and siennas, and like everything else are semi-transparent, enabling the immersant to see through them and float through them as well. As the immersant looks around, she may see the river again but manifested here as a swiftly flowing underground stream, a luminous green ribbon suspended in space, issuing its own distinctive sounds. If she approaches, its gravitational pull will seize her as did the forest river, and carry her along, while summoning in either the forest or the body realm below. Perhaps she will surrender to the underground stream (for it takes bodily effort to move away from its pull) and find herself back in the forest landscape, its various abstractions now engaged in the lushness of summer green or the ochres and siennas of autumn.

FIGURE 4.23 Autumn forest landscape, *Ephémère*, 1998, digital image captured in real-time through head-mounted display during live immersive journey/performance.

The immersant may, however, choose to remain longer in the under-earth, seeking the inter-responsive 'seeds'. Suspended among the subterranean rocks, at a certain depth, and appearing only at certain times within the temporal progression of the work, are pod-like seeds (which like all other elements in *Ephémère*, are engaged in visibly and audibly coming into being, lingering and passing away).

If the immersant gazes at a seed from a certain distance, it will begin to flicker in acknowledgment that it has 'sensed' her presence, and will consequently begin to germinate. If the immersant is able to approach before it completes this process and withdraws, she can enter inside and experience its luminous blooming while enveloped within its translucent veils. After the seed reaches its climax, it will fade and she will find herself alone again in the subterranean realm. Some immersants never notice the seeds, some do not gaze long enough to initiate their germination, while others effortlessly enter their blooming.

Below the subterranean earth is yet another realm, the interior body, intended to suggest the internal pulsing frothing rhythms, aural and visual, of a subjectively inhabited body engaged in the processes of living (referring not only to the immersant's own body, but to all living flesh bodies). This realm has been placed at the 'bottom' of *Ephémère* to function metaphorically as the substrate or foundation, under the fecund earth and the lush bloomings and witherings of the land. Recalling the words of Joseph Campbell in *The Inner Reaches of Outer Space: Metaphor as Myth and Religion* (1986): 'Myths and

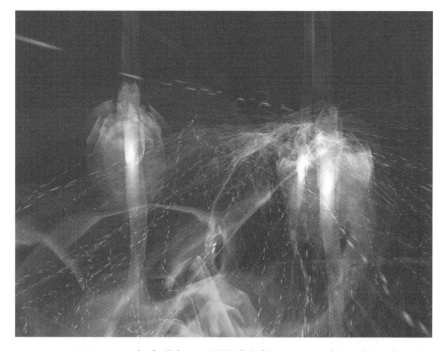

FIGURE 4.24 Seeds, *Ephémère*, 1998, digital image captured in real-time through head-mounted display during live immersive journey/performance. See also colour insert, Plate 5.

dreams . . . are motivated from a single source – namely the human imagination moved by the conflicting urgencies of the organs.'

As in the other realms of *Ephémère*, here all organ, artery and vein-like elements are semi-transparent and semi-abstract, and are engaged in carefully choreographed yet random appearings and disappearings. If the immersant floats to a certain depth within the body realm, she may be surrounded by a slow raining of blazing white egg-like forms, which eventually drift by and fade away. In this realm too, there is the river, but manifested here as a bloodstream or artery from which issue forth the sounds of calling voices. Sampled from recorded animal cries – the only non-viola-derived sounds in the work – these are intended to reaffirm, subliminally, embodied presence.

From within the body, the immersant can descend even lower, passing through a transitional realm of thorn-like forms whose colour shifts from blood red to green as she is returned to the upper foliage of the forest landscape.

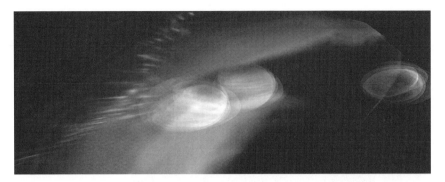

FIGURE 4.25 Body (Egg), *Ephémère*, 1998, digital image captured in real-time through head-mounted display during live immersive journey/performance.

Similarly, if the immersant rises steadily in the forest, she will pass through the thorns until she is again in the body. In this way, it is possible to cycle endlessly vertically through *Ephémère*: if the immersant attempts to seek its horizontal limits her quest will reveal that each realm is endless as its elements constantly reconfigure themselves around her.

There are several 'endings' in *Ephémère*, all dependent on where the immersant is during the last phase of her journey. If within the forest landscape a dozen minutes into the work, she will be enveloped by brightly-hued orange space with dark tree ghosts screaming distantly and falling all around (again, the *Osmose* tree). If she chooses to remain within this autumnal burning, eventually all that will remain are single ochre leaves falling in dark space. If she is within the under-earth at this late phase, she will experience a distinctive shift in its sounds as its various elements begin disappearing one by one, until there are only embers falling. And if she is within the body realm, she will experience the flesh around her slowly giving way to luminous filaments of bone which in turn will also fade out, leaving only embers and ash drifting through space, ending the journey in dissolution.

The immersive experience

The immersive experience of *Osmose* and *Ephémère* is designed to be intimate and solitary. During public exhibitions, however, the experience takes on a performative aspect. In this context, the immersion chamber is located adjacent to a large dark space where visitors assemble. Here, the immersant's journey

FIGURE 4.26 An immersant in *Osmose*, seen through the shadow silhouette screen. See also colour insert, Plate 6.

is projected on a wall in real-time, i.e. as it is being experienced live by the immersant. This space is also filled with the sounds being generated by the immersant's behaviour. In addition, the shadow silhouette of the immersant's body is cast on another wall as he/she moves and gestures within the work. The use of this shadow-silhouette alongside the real-time projection is intended to draw attention to the body's role as ground and medium for the experience.

Since 1995, more than 20 000 people have been individually immersed in *Osmose* and *Ephémère*. We have had an opportunity to observe many of them and have noticed certain patterns of behaviour. After becoming accustomed to using the interface of breath and balance, most participants are first intent on 'doing', i.e. travelling around at high speed to see as much as possible, in what appears to be an extension of everyday goal-oriented, action-based behaviour. However, half way into the fifteen-minute experience, most people undergo a change: their facial expressions and bodily gestures loosen, and instead of rushing, they begin to slow down, as if perceptually mesmerised. In this final phase, attention seems to be increasingly directed toward the unusual sensations of floating and seeing through things, in what becomes a kind of slow-motion perceptual free-fall.

Based on participant responses gathered through written comments, correspondence and video interviews, it appears that many people experience a heightened awareness of self-presence – paradoxically consisting of both a sense of freedom from their physical bodies and a heightened awareness of being in their bodies at the same time. Often, people experience intense feelings of euphoria and/or loss when the session is ending, causing some participants to cry afterwards and others to even exclaim they are no longer afraid of dying. What is going on here? As I have suggested in earlier essays (see 'Further Reading'), a partial answer may lie in the words of the French philosopher Gaston Bachelard:

> By changing space,
> by leaving the space of one's usual sensibilities,
> one enters into communication with a space
> that is psychically innovating . . .
> For we do not change place, we change our nature.
> Gaston Bachelard, *The Poetics of Space*

Bachelard was actually referring to the perceptually and psychologically transformative potential of places like the desert, the plains and the deep sea – immense open spaces which are perceptually invigorating because they are *unlike* the environments to which most of us are accustomed. Bachelard's insight has been echoed by psychologists researching the effects of traditional methods of achieving altered mental states. In such practices as meditation (which involves deep breathing as do *Osmose* and *Ephémère*), as well as chanting, dance, and the ingestion of psychotropic plants, the intent is to foster psychological conditions which lead to an 'undoing' of habitual perceptions in favour of alternative sensibilities. While these may be less efficient in terms of biological survival, psychologists believe that they permit experience of aspects of reality previously ignored. According to Arthur Deikman in 'Deautomatization and the mystic experience' (1990), the conditions fostered by such practices involve a de-habituating or 'de-automatizing' of perceptual sensibilities, which leads to perceptual expansion.

Conclusion

If the responses of those who have experienced the environments *Osmose* and *Ephémère* are anything to go by, then it appears that immersion in virtual space can be 'psychically innovating'. That this may be so, is, I believe, due to the paradoxical nature of the medium. Here, ephemeral virtuality can coexist with an apparent 'real' three-dimensionality of form, and feelings of disembodiment can coexist with those of embodiment (given the use of an embodying interface). These experiential paradoxes, when combined with the ability to interact kinesthetically with elements within the space, can create a very unusual perceptual context, providing a unique means of 'changing space'.

I want to emphasise, however, that the medium's perceptually refreshing potential is possible *only* to the extent that the virtual environment is designed to be *unlike* those of our everyday experience. When designed in ways that merely reflect our habitual perceptions and culturally biased assumptions, such environments forgo their transformative potential and serve to reinforce King Logos and the status quo. It is only when such environments are constructed in ways that circumvent or subvert the medium's conventions that immersive virtual space can be used to convey alternative sensibilities and worldviews, potentially functioning as a perceptually and conceptually invigorating philosophical tool.

At the present time, I am beginning a new work in immersive virtual space, and consider *Osmose* and *Ephémère* as only early steps in what I hope to accomplish with this medium. Ultimately, what I am seeking are even more effective ways to use this technology to provide an experience for others whereby it is possible, however momentarily, to slip through Aldous Huxley's 'doors of perception' and glimpse reality, as I have intuitively sensed it, and so deeply long to know it, beyond the Cartesian divide.

FURTHER READING

Bachelard, G., *The Poetics of Space*, Boston: Beacon Press, 1966.

Campbell, J., *The Inner Reaches of Outer Space: Metaphor as Myth and Religion*, New York: Harper Collins, 1986.

Coyne, R., 'Heidegger and virtual reality: the implications of Heidegger's thinking for computer representations', *Leonardo: Journal of the International Society for the Arts, Sciences and Technology*, **27** (1994), No. 1.

Davies, C. 'Osmose: notes on being in immersive virtual space', *Digital Creativity*, **9** (1995), No. 2, 65–74.

Davies, C., 'Changing space: VR as an arena of being', in *The Virtual Dimension: Architecture, Representation and Crash Culture*, ed. John Beckman, Boston: Princeton Architectural Press, 1998.

Davies, C. 'Landscape, Earth, Body, Being, Space and Time in the Immersive Virtual Environments Osmose and Ephémère', in *Women, Art and Technology*, ed. J. Malloy, Boston: MIT Press, 2002.

Davies, C. and Harrison, J., 'Osmose: towards broadening the aesthetics of virtual reality', *Computer Graphics: Virtual Reality*, **30** (1996), No. 4.

Deikman, A., 'Deautomatization and the mystic experience', in *Altered States of Consciousness*, ed. C. Tart, New York: Harper Collins, 1990.

Gigliotti, C., 'Aesthetics of a virtual world', *Leonardo: Journal of the International Society for the Arts, Sciences and Technology*, **28** (1995), No. 4.

Hayles, N. K., 'The seductions of cyberspace', in *Rethinking Technologies*, ed. V. Conley, Minneapolis: University of Minnesota Press, 1993.

Hayles, N. K., 'Embodied virtuality: or how to put bodies back into the picture', in *Immersed in Technology: Art and Virtual Environments*, ed. M. Moser, Cambridge: MIT Press, 1996.

Huxley, A., *Doors of Perception*, New York: Harper & Row, 1954.

Jones, R., *Physics as Metaphor*, Minneapolis: University of Minnesota Press, 1982.

Leder, D., *The Absent Body*, Chicago: University of Chicago Press, 1990.

Lefebvre, H., *The Production of Space*, Oxford: Blackwell, 1991.

Merleau-Ponty, M., *The Phenomenology of Perception*, London: Routledge & Kegan Paul, 1962.

Merleau-Ponty, M., 'Eye and mind', in *The Primacy of Perception*, Chicago: Northwestern University Press, 1964.

Merleau-Ponty, M., *Signs*, Chicago: Northwestern University Press, 1964.

Sardar, Z., 'alt.civilizations.faq: cyberspace as the darker side of the West', in *Cyberfutures: Culture and Politics on the Information Superhighway*, ed. Z. Sardar and J. Ravetz, London: Pluto Press, 1996.

5 Mapping space

LISA JARDINE

A map is not a neutral representation of the relative positions of places in space. Even something as apparently innocent as a city map is selective about what it represents. Here I am concerned with maps from one of the great ages of mapping, the Renaissance, and with the agendas embedded within those maps. A Renaissance map typically records the aspirations of the person for whom it was designed. The maps that I will discuss were designed for an emperor, Charles V, and his dynasty. I will dwell especially on some surprisingly detailed maps incorporated into several series of large, commemorative tapestries. These extraordinary art-works travelled throughout Europe in the sixteenth century, objects of wonder which made vivid the political supremacy of their owner. As we explore the meanings of these objects, and the material and social conditions that made them possible, we enrich our understanding of maps as the instruments of power, and also edge toward a new understanding of the Renaissance itself, seeing it less as the flowering of a uniquely European spirit than as the joint production of Europe and its oft-mapped neighbours to the east.

A map of the world: the 'Spheres' tapestries

In 1525, Catherine of Austria married John III of Portugal, thereby uniting two of the most powerful royal houses in sixteenth-century Europe – the House of Habsburg, headed by Catherine's brother, the Holy Roman Emperor and King of Castile, Charles V, with the Portuguese House of Avis, headed by John III, self-styled 'Lord of Guinea and of the Conquest, Navigation and Commerce of Ethiopia, Arabia, Persia and India'. Costly luxury goods accumulated around the royal wedding. As befitted a Habsburg princess, Catherine brought

Space: In Science, Art and Society, edited by F. Penz, G. Radick and R. Howell.
Published by Cambridge University Press. © Darwin College 2004.

105

with her to Lisbon an impressive collection of tapestries, together with illuminated manuscripts, cartloads of family jewels, ornate household furniture and a magnificent wardrobe of the most stylish and expensive garments in her trousseau.

As part of the lavish expenditure associated with the wedding and its accompanying gifts and ceremonies, a series of lavish tapestries was commissioned, in all likelihood by Catherine's Habsburg family. Entitled 'The Spheres', this tapestry series was designed by the Flemish painter Bernard van Orley (or is generally attributed to him), and executed by the world-renowned tapestry workshops in Brussels.[1] The series commemorated a great and glorious union centred on, and celebrating, Portugal's ruling dynasty. Its display of Renaissance 'magnificence' was carefully designed to emphasise the political importance of the bride and the Habsburg dynasty, the groom and the reach of the Portuguese maritime empire, and the extent of the awesome territorial alliance achieved by this landmark marriage. By grounding their (to be brutal) forceful propaganda message in graphic beauty, rather than choosing to commemorate the event with ceremonial armour, swords and other paraphernalia of war, as they might equally well have done, the families of the couple also announced emphatically how civilised and humane, in spite of the political imperatives, the imperial Habsburg–Avis liaison of which they formed the beginning would be.[2]

The third (and central) tapestry in this series of three is entitled 'Earth Under the Protection of Jupiter and Juno'. At its centre is a terrestrial globe, flanked by the monumental figures of Jupiter and Juno – or perhaps John III and Catherine as Jupiter and Juno (see Fig. 5.1). The inscription above is 'Gloria summa, / Nam sua ipsius sola' – 'His glory is the greatest, / For it is his alone'. Abundance and wisdom hover over Jupiter, fame and renown or victory over Juno. Above the globe is a brilliantly shining sun; below it a radiant moon.

---

[1] Throughout the period I shall be concerned with here, Brussels was under Spanish rule – as the Belgians saw it, under hostile occupation. For the relationship between the Spanish occupation of the Low Countries and emerging northern European humanism and its ideals, see L. Jardine, *Erasmus: The Education of a Christian Prince* (Cambridge: Cambridge University Press, 1997).

[2] On the 'Spheres' tapestries see Jerry Brotton, *Trading Territories: Mapping the Early Modern World* (London: Reaktion Books, 1997), 17–19; A. D. Ortiz (ed.), *Resplendence of the Spanish Monarchy: Renaissance Tapestries and Armor from the Patrimonio Nacional* (New York: The Metropolitan Museum of Art, 1991), 55–67; H. Soly and Johan van de Wiele (eds.), *Carolus: Charles Quint 1500–1558* (Ghent: Snoeck-Ducaju & Zoon, 2000), 284–5.

FIGURE 5.1 Third tapestry of the 'Spheres' series of three: 'Earth Under
the Protection of Jupiter and Juno', attributed to Bernard van Orley, Brussels,
c. 1520–30, gold, silver, silk and wool, 11ft 2ins by 11ft 4ins, © Patrimonio Nacional,
Madrid; next page, detail. See also colour insert, Plate 7.

Jupiter and Juno are supported by mythic figures, perhaps representing the
constellations behind them.

The winds which carry Portugal's trading ships around the known world
frame the brilliantly executed globe. For it is the mapped portion of the globe,
facing the viewer, which is surely intended as the triumphalist centre of the
composition. It shows the entire continent of Africa, and, to the east, India and
the 'Spice Islands', the Moluccas. At carefully spaced intervals along the coast
of Africa and India, and on the islands of the East Indies, are planted flags
bearing the arms of Portugal, proclaiming these territories as theirs. Tucked

FIGURE 5.1 (*cont.*). See also colour insert, Plate 8.

up at the top is Europe. John III holds his sceptre over Portugal, while the Habsburg Catherine gestures toward the whole region. Their overseas territorial possessions are brightly coloured in orange and yellow, and scattered with images of exotic animals and curious savages. Europe is discreetly muted in its tones.

A year after the marriage between Catherine and John, in 1526, with perfect symmetry, and as if to give total finality to the Habsburg–Portuguese dynastic union, Charles V married Isabella of Portugal, John III's sister, in Seville. The children of these two marriages were as close relatives as it is possible, without incest, to be – first cousins, with just two sets of grandparents between them.[3]

Although the 'Spheres' tapestries were undertaken in the year of Catherine and John's marriage, their intricacy, and the ambitiousness of their design, meant that they were not completed until long after the wedding ceremonies. The first record we have of their public display, indeed, is one entire generation later, to mark a further Portuguese–Spanish consolidating match. This time it was the wedding, in 1543, of the sixteen-year-old daughter of John III and Catherine, Maria, to Charles V and Isabella's sixteen-year-old son Philip. Given the nearness in blood of the young couple, the imagery of the tapestry sequence was no less appropriate for this wedding than for the one for which it had originally been intended.[4] The Portuguese Maria and Habsburg Philip were first cousins – recall that Catherine and Charles V were sister and brother, and so were their royal partners, John III and Isabella – so in any interpretation of the 'message' of the tapestry to admiring onlookers, it would have been simple enough to reverse the original allocations of metaphorical meaning, and now to claim that Jupiter represented the Habsburg Philip, and Juno the Portuguese Maria. On the updated reading, it was Philip's sceptre that now firmly laid claim to Lisbon for his dynasty, while Maria gestured toward the Portuguese maritime empire she brought him as her dowry.[5]

As a statement of imperial aspiration, the 'Spheres' tapestries proved predictably prophetic. In 1580, Habsburg Philip, now Philip II of Spain, became King of Portugal. In the words of the historian of the Habsburgs, Jean Bérenger:

---

[3] In fact, under sixteenth-century canon law, Charles' marriage to his sister's husband's sister was incestuous by affinity (as opposed to blood incest).

[4] Marriage between first cousins (let alone first cousins of brother–sister marriages) was prohibited by the Catholic Church, but the Habsburgs specialised in such close alliances.

[5] Between these Habsburg-initiated marriages, strengthening Charles V's claims on Portugal and his political ambitions for a pan-European power bloc there was in fact one other, incorporating the French royal dynasty alongside the Portuguese and Austro-Spanish (Habsburg). In 1530 Charles V's older sister, Eleanor, second wife (now widow) of King Manuel of Portugal, and step-mother to John III (her sister's husband), was married again, this time to the French King, Francis I.

> Both [Portugal and the Habsburgs] benefited from the arrangement. The Habsburgs placed under their authority the Portuguese colonies and trading stations in Asia, Africa and America; they thus had control of the entire world economy, the production of precious metals and the lucrative trade in spices, and put together the most powerful war fleet and merchant army. The Portuguese for their part had easier access to the American silver which they needed to pay for their purchases in the Far East; they still purchased more oriental goods than they sold European products and settled the deficit in their balance of payments with the massive export of precious metals. In addition, Philip II for the first time realized an old dream – the unification of the Iberian peninsula under one crown.[6]

Amid all this power-brokering and dynastic manoeuvring, this compelling graphic representation, in an unfamiliar medium, invites a series of questions: What kind of an art object is this cartographic tapestry? What is its relationship to that flowering of art and culture we continue to refer to as the Renaissance? Does the robustly political context in which the 'Spheres' tapestries were ostentatiously shown off have implications for that 'flowering'? These are the questions I shall be trying to answer in what follows.

East and west: the 'Los Honores' tapestries

It will help to set the 'Spheres' tapestries beside another territorially-aware tapestry series from the same era, indeed from that same glorious union of the Houses of Habsburg and Avis which the 'Spheres' tapestries both celebrate and prefigure. As we have noted, the Catherine – John III wedding was the first part of a double-wedding contract of a kind the Habsburgs particularly favoured (and put repeatedly into practice) as the optimal way of cementing important dynastic alliances.[7] The year following the marriage of Catherine and John III, the Emperor Charles V himself married Isabella of Portugal. For the ceremony, all the relevant spaces were decorated with tapestries from the Emperor's existing large store of magnificent series. Afterwards, Charles and

---

[6] J. Bérenger, *A History of the Habsburg Empire 1273–1700* (London: Longman, 1994), 215.

[7] See, for example, the account of Maximilian's paired wedding contracts uniting the Habsburg line with the Jagiellon ruling house of Hungary, in L. Jardine, *Worldly Goods: A New History of the Renaissance* (London: Macmillan, 1996), 277–83. Here, too, substantial loans from the Emperor's Fugger bankers eased the negotiations.

his new wife were presented with a gargantuan nine-piece set of new tapestries called 'Los Honores'.

It is important for my argument here to appreciate the imposing scale of the tapestries discussed. These are works of art to daunt and awe the onlooker. 'Earth Under the Protection of Jupiter and Juno' is almost 12 feet square; in other words, the figures flanking the globe are considerably larger than life-size. Each of the 'Los Honores' tapestries is over 28 feet long and 16 feet high – the size of the side of a double-decker bus. Their impact on those who encountered them, displayed along the corridors of power, cannot be overestimated (see Fig. 5.2).

The commission for the 'Los Honores' tapestries dated back to 1520, the year in which Charles was crowned 'King of the Romans' and Emperor Elect at Aachen.[8] In 1522 Pieter Coeck van Aelst, a tapestry weaver and dealer, and later tapestry keeper for Charles, borrowed a large sum from the Fugger bankers via their Brussels agent to begin making the tapestries, whose value he put at 3 050 Flemish pounds. By 1523 the Fuggers were holding seven of the tapestries as security. In 1525 van Aelst proposed to the bankers that the series as a whole should be offered for purchase to the Emperor himself. When I say that the 'Los Honores' tapestries were 'presented' to Charles V at his wedding in 1526, I actually mean that the fifth tapestry of the series ('Honour') was unveiled to him with suitable ceremony by the dealer as a 'taster' for the as-yet-incomplete series, and Charles was invited to purchase the complete set of nine.[9] In the event Charles agreed, and the sum of 12 000 Spanish ducats was added to his outstanding debts to the Fuggers, relieving van Aelst of his

---

[8] 'On the tapestry with the allegory of *Fortuna*, which opens the set, the year 1520 is woven into a cartouche on the left-hand side of the small platform on the right. Fortune's wheel itself bears at the top the imperial crown, the sceptre and the sword of state of the Holy Roman Empire . . . The *Fortuna* tapestry, and with it the whole set, originally called after Fortune, are in fact a congratulation and exhortation for the young emperor' (G. Delmarcel, *Los Honores: Flemish Tapestries for the Emperor Charles V* (Antwerp: Pandora, 2000), 16).

[9] According to surviving documents, had Charles turned the set down, it would then have been sold elsewhere to the highest bidder: 'Pieter van Aelst urged that the whole set of nine *Fortuna* tapestries should be offered to the emperor, for whom they had been made, and who therefore had a prior claim to the set, "except, however, that he, Haller [the Fugger agent] should first and foremost offer them to his imperial majesty for whom the tapestries – as he, Pieter, declared – were made, will have to inform him and graciously request whether he would be interested in buying and would wish to receive them and pay for them or not"' (Delmarcel, *Los Honores*, 9).

FIGURE 5.2 Left, left side of the 'Fortuna' tapestry from the 'Los Honores' series, by Pieter van Aelst; right, right side of the 'Fortuna' tapestry. © Patrimonio Nacional, Madrid.

FIGURE 5.2 (cont.).

own debt burden.[10] Two contemporary chronicles provide us with documentary evidence that the 'Los Honores' tapestries were displayed in Valladolid a year later, in the church where Charles and Isabella's new-born son Philip was christened, and along the processional route down which the baby prince was carried.[11]

In sum, we have a series of magnificent marriages (and a christening), and a collection of fabulously expensive tapestries – priceless objects of exquisite beauty and imposing presence, which decorated the locations for these events, while at the same time conveying strong messages about the imperial power-rating of their owners. Moreover, just as the 'Spheres' tapestry advertised the extent of that power by centring on the space of Africa, India and the Moluccas, the 'Los Honores' tapestries are filled with exotic depictions of figures from myth, scripture and history as strangers from distant lands and of other faiths.

Not only does 'Mahomet' – in full Ottoman dress – appear in two of the tapestries, but figures from antiquity such as Julius Caesar, or from the Old Testament such as David (both represented in the 'Fama' tapestry in the sequence), are robustly oriental. In these extraordinarily mobile tapestry depictions, pulsating with life and action, figures from eastern and western history and legend, sacred scripture and mythology, confront one another in encounters of almost cinematographic intensity. In the complexly narrated moral tales which shape the tapestries' compositions, recognisably European heroes and heroines (including Charles V himself) triumph over a vivid array of visitors from the sixteenth-century global village.[12]

Art, power, space

The 'Los Honores' tapestries celebrate the imperial power of the young Emperor, and, as the titles of individual tapestries make clear, exhort him to glory ('Honour', 'Fortune', 'Fame', 'Nobility'), virtuous conduct ('Virtue', 'Prudence', 'Justice'), and the defence of Christendom against the infidel ('Faith',

---

[10] For the whole story see Delmarcel, *Los Honores*, 9–10.
[11] See Demarcel, *Los Honores*, 10–11.
[12] It is interesting to set alongside such 'exotic' versions of cultural competition early sixteenth-century versions of the Greek philosophers Plato and Aristotle, who are comfortably treated graphically as exotic orientals.

'Infamy'). It is important to notice how this series, like the 'Spheres' series, directs its attention eastwards, visually matching the awesome aspirational might of its Habsburg reach against exotic images from competing imperial centres elsewhere – this time, the equal and opposite global dominion of the Ottoman Turks under Suleiman the Magnificent.[13]

The apparently innocuous globe in the third 'Spheres' tapestry superimposes its cultural claims – to ultimate beauty, to unmatchable aesthetic extravagance – upon the international political power map. Renaissance art is spatially co-extensive with territorial ambition. Taste and beauty fuse, in the 'Jupiter and Juno' tapestry, around the contested space of the rich commercial world to Europe's east. The mythologies of Roman antiquity frame the arena of imperial power politics. The globe between Jupiter and Juno is less a representation of the Earth than a participant in an encounter between 'civility' and economic and political competition.

Much the same blend of aesthetic, cartographic and political elements, to much the same purpose, can be seen in another globe, the Habsburg Globe constructed by the great cartographer Mercator for Charles V in 1541 (see Fig. 5.3). Here is no simple navigator's or traveller's tool. This is a piece of décor for the Emperor's study – an accessory to rule, on which the ruler can imperiously lay his hand (or indeed, like Jupiter, his sceptre), and point out the reach of his dominion to visiting ambassadors.

This globe is copied, with up-to-the minute modifications, from a previous one made for Charles in 1536 by Mercator's teacher, the great cartographer Gemma Frisius.[14] Charles had sponsored the production of a globe to commemorate his 1535 military triumph over the Ottomans at the siege of Tunis. In text and image, like the prominent Habsburg arms above the cartouche, it lays claim to the territories mapped onto its surface. At Tunis, the Emperor's troops took back the North African trade centre from Suleiman's forces, and sacked the town with all the ruthlessness generally attributed to the Turk. On Gemma Frisius' globe, a flag bearing the Habsburg ensign, the imperial eagle,

---

[13] For a full general account of the importance of the Ottomans in the shaping of the European Renaissance aesthetic see Jardine, *Worldly Goods*. For a recent, more detailed account see J. Brotton and L. Jardine, *Global Interests: Renaissance Art between East and West* (London: Reaktion, 2000).

[14] Robert Haardt, 'The Globe of Gemma Frisius', *Imago Mundi*, **9** (1952), 109–10.

FIGURE 5.3 Mercator globe, copied from one by Gemma Frisius. The Frisius
globe is reproduced in the introductory essay to Soly and van der Wele (2000);
see note 2. Reproduced with permission of the Fürst Thurn und Taxis Zentralarchiv,
Regensburg, Germany.

flies over the captured city.[15] Although a minor campaign, in which the Habsburg victory was so assured from the outset that Charles took war artists with
him to record his triumph graphically for posterity, the conquest of Tunis nevertheless marked a symbolic stand against the westward encroachment of the

[15] On the political, commercial and territorial importance of Frisius' and Mercator's
globes see Brotton, *Trading Territories*, 155–60.

Ottoman Empire. It became, as we shall see, an iconic reference point in the supposed global supremacy of the Habsburgs.[16]

Like the globes of Frisius and Mercator, the 'Spheres' and 'Los Honores' tapestries were simultaneously status symbols and works of art. They were hugely costly (in terms of materials and the skilled labour and time that went into producing them), and, once completed, they became, in their near-uniqueness, almost priceless.[17] Conveniently portable when rolled, the tapestries travelled with the imperial households, transforming each stopping point along a progress into appropriately sumptuous accommodation. They shaped the aesthetic of the court – in the court circle, panel paintings were regarded as cheap and inferior substitutes for wall-to-wall hangings. Displayed in the rooms where the Emperor received everyone from ambassadors to visiting royalty, the tapestries bordered those encounters, becoming an intrinsic part of the experience of being in the presence of imperial power. As we learn from surviving letters and reports, both the aesthetic and the 'messages' of the representations left a lasting impression.[18] This is the culture which formed the Renaissance mind.

A historiographic interlude

For all their visibility in the Renaissance, tapestries like these have become virtually invisible to historians of the Renaissance. To understand why, we have to return, briefly, to that moment in the late nineteenth century when the very term 'Renaissance' was coined in the sense in which we all continue to use it. Michelet in France, Burckhardt in Germany and Walter Pater in England collectively argued that a 'reflowering' of art and literature had occurred in

---

[16] Tunis had once been Carthage (see Gonzalo's comment in Shakespeare's *Tempest*, Act 2, scene 1, line 80: 'This Tunis, sir, was Carthage'). In conquering Tunis, Charles V could therefore readily compare himself to Scipio triumphing over Hannibal in the second Punic war – a role Francis I of France had already adopted for himself in commissioning the 'Triumph of Scipio' tapestry series. See Brotton and Jardine, *Global Interests*.

[17] Although several series might be produced of the same cartoons, both the *editio princeps* and the 'copy' or 'copies' were almost equally objects of rarity, available only to the richest in society.

[18] In 2000, the nine fully restored 'Los Honores' tapestries went on show in Mechelen, Belgium, as part of the celebrations of the five-hundredth anniversary of the birth of Charles V. The impact of the tapestries, which were hung in purpose-built spaces, where each tapestry entirely covered one wall of the three rooms in which they were displayed, was extraordinary – an experience which was almost cinematographic, such was the force of the more-than-life-size figures when seen from up close.

the fifteenth and early sixteenth centuries, and that the groundplan for a pecu-
liarly European mind, and an associated aesthetic ideal, were thus brought into
being.

It was Jacob Burckhardt's great book, *The Civilization of the Renaissance
in Italy*, first published in German in 1860, which directed historians to the
fifteenth and sixteenth centuries in Europe as the matrix of a recognisable,
and distinctively European, modernity. As Burckhardt once explained, in a
letter describing the purpose of his literally epoch-making book, he had set out
to show both that there had been a rebirth of art and thought in this era, and
that this rebirth or, in French, 'renaissance', was 'the mother and the source of
modern man'. His heavily moralised narrative argued that an antique tradition,
retrieved by scholars of humane learning in Italy in the fifteenth century, gave
rise to an idealised and aesthetically pure fine art, a delight in truth and beauty,
and, above all, to a robust individualism and a new intellectual freedom. These,
he claimed, were enduringly the cultural triumphs of the West, distinguishing
western civilisation permanently from the barbarism of both the Middle Ages
and the cultures to the east of the Mediterranean.

Burckhardt's argument had a corollary. The civilised values handed down
from antiquity had to be retrieved from amid the social and intellectual debris
of late medieval barbarism. In order for this to happen, the Renaissance individual had for the first time to recognise his identity and urges as separate
from visible civic forms, and to shape himself consciously to the new social
world he aspired to inhabit. The cost of progress had to be strenuous personal
self-control. For Burckhardt, the civilised values of the Renaissance required
that men limit their natural drives in the interests of an urbane mode of living,
or humanity. The anxiety that such a curtailment of instinctual life generated
was, Burckhardt argued, an inevitable accompaniment of the civilised temper-
ament. The removing from view – away from the aesthetic gaze of Europe – of
the art and thought of the Orient was an essential condition for civilisation as
we know it.[19]

---

[19] Although the word 'Renaissance' is to be found in the seventeenth century, the use
of the term to describe this re-emergence of the high culture of antiquity in western
Europe derives from the nineteenth century. Burckhardt's *Die Kultur der Renaissance
in Italien* was first published in 1860; it was republished with an important new
introduction in 1930. Jules Michelet is supposed to have been the first modern historian to apply the comprehensive term 'Renaissance' to the whole epoch; but though
Burckhardt did not coin it, it was he who made it popular, and his essay was the
starting point for all the subsequent discussions about the beginning and the concept

Tapestry series do not fit well into Burkhardt's schematic. They lack both the individualised creative originality – the sustained presence of the artist of genius, forever present in the brush strokes on canvas – and the absolute uniqueness of the great panel paintings and canvases which Burckhardt so admired. With their team-work execution, combining the skills of an artist and armies of skilled designers and weavers, not to mention the commercial agents responsible for acquiring and underwriting the costs of the materials, and the assumption that there will be multiple copies of the 'original' series, they fail to fit the nineteenth-century criteria for 'Renaissance art works' out of which the aesthetic of the modern allegedly was born. As a result, these extraordinary objects have languished in museums and private collections, invisible in plain sight, inconsequential cladding for stairwells and the walls of entry-halls.

Let us continue our exploration of these important clues to relations between art, power and space in the Renaissance. We return to the imperial Habsburgs, and in particular the conquest of Tunis, memorialised in Frisius' globe and in one of the most remarkable Renaissance tapestries of all. It is to the events behind those tapestries that we now turn.

Before Tunis: tapestries in the rivalry between the Ottoman East and the Habsburg West

The run-up to the 'Conquest of Tunis' tapestries begins on the night of 23–4 February 1525, when the imperial troops of Charles V joined battle with those of the French sovereign, Francis I, just outside Pavia. Two hours later the French king had been captured and his troops had surrendered. The defeat of the French had long-term political consequences for the whole of Europe, giving the Habsburgs the upper hand over the French for the control of Italy.[20]

In 1531 Charles V was presented with a series of seven magnificent tapestries commemorating the humiliating defeat and capture of the French king at Pavia.[21] The artist responsible for the designs was, once again, Bernard van

---

of the Renaissance. See R. Pfeiffer, *The History of Classical Scholarship from 1300 to 1850* (Oxford: Clarendon Press, 1976), 18.

[20] The marriage of Charles V's widowed sister Eleanor to Francis I in 1530 was part of the peace negotiations between the victorious Charles and the vanquished Francis.

[21] On the 'Battle of Pavia' tapestry series, and the military campaign it recorded, see N. Spinosa *et al.*, 'To the Glory of Charles V', *FMR*, **106** (October/November 2000),

Orley.[22] The series came from the prestigious Van der Moeyen tapestry works in Brussels, and was perhaps commissioned by Charles' third sister, and Regent in the Low Countries, Mary of Hungary. A complete set of drawings for these tapestries survives in the Louvre.[23]

The 'Battle of Pavia' tapestries broke new ground in the propaganda and conspicious-consumption contest amongst the imperial rulers of Europe and beyond. The events represented are entirely 'real' – commemorations, with graphic documentary impact, of a great and significant victory, stripped of the residual antique mythology and moralising of the 'Los Honores' tapestries.[24] In the central tapestry of the series, Francis quite literally 'falls' – unhorsed and humiliated on the battlefield (see Fig. 5.4). The military action is surrounded by scenes and figures 'from life', precise down to the least detail of clothing and accessories.

With such a virtuoso series of tapestries now to its credit, the Van der Moeyen firm looked around for prestigious commissions from other imperial rulers eager to keep up with the Habsburgs. Like the Habsburgs themselves, the firm looked east for expansion and new clients. In 1533, they sent Pieter Coeck van Aelst, the dealer responsible for the 'Los Honores' tapestries, as their agent to the court of the Ottoman Sultan Suleiman the Magnificent in Constantinople, with the aim of interesting Suleiman in becoming a customer, and commissioning his own tapestry series.[25] Apparently such a series was initiated,

70–110; L. Casali *et al.*, *Gli arazzi della battaglia di Pavia nel Museo di Capodimonte a Napoli* (Pavia: Vigi Effe, 1993).

[22] Few of these drawings for tapestry sets survive. When they do, they amaze art historians for their aesthetic brilliance (a brilliance the same historians have trouble seeing in the finished tapestries). As a result, scholars have been inclined to assign those that do survive (like these exquisite 'Battle of Pavia' drawings) to a known 'artist', such as van Orley, rather than to a 'jobbing tapestry designer' like Pieter Coeck van Aelst. That said, there is some evidence that van Aelst was the designer (see note 23).

[23] A set of woodcuts by van Aelst of the Battle of Pavia suggests that van Aelst and Orley may have competed for the commission.

[24] The change is even more striking if one compares the various sequences of tapestries showing the triumphs of Alexander, Scipio, and so on, made for the various imperial heads of houses, and in which the ancient prototype figures the modern emperor or monarch.

[25] In the same year, the Austrian merchant Rechlinger bought sample tapestries from the Van der Moeyen firm of the seven-strong 'Battle of Pavia' series and the twelve-strong 'Hunts of Maximilian' series (both designed by Bernard van Orley, and commissioned by Mary of Hungary). Rechlinger's Venetian partner Marco di Niccolò joined van Aelst in Constantinople with these samples to show Suleiman.

FIGURE 5.4 Detail from the 'Battle of Pavia' tapestry, 'Sortie of the besieged'; right, detail from the 'Battle of Pavia' tapestry, 'Capture of the King of France'. Reproduced with the permission of the Museo Nazionale di Capodimonte.

FIGURE 5.4 (*cont.*)

though never completed.[26] In van Aelst, the imperial art contest found its ideal representative, at once a generator of competition between rival imperial patrons and collectors – he showed the Sultan specimens of Habsburg tapestries under production – and an exemplar of the creative brilliance of European artists, capable of producing skilful designs for the projected Suleiman series.[27]

Van Aelst spent a year in Constantinople, and became, apparently, a great favourite of the Sultan. At the end of the year, we are told, he returned to Europe a wealthy man; 'by the royal bounty of Suleiman's own hand', he was dismissed with honourable gifts, a ring, a jewel, horses, robes, gold and servants, which at Brussels he converted into an annual pension. He also turned his drawings of the Sultan and his court, for the unexecuted tapestries, into a series of woodcuts, which he published extremely successfully in Antwerp in 1553, under the title *The Customs and Habits of the Turks*.[28]

The speculative commercial trip to Constantinople, undertaken and underwritten by the Brussels tapestry dealers, provides us with evidence of deliberate exploitation by the tapestry dealers of potential competition between imperial patrons.[29] The expectation of a successful marketing coup rested on a confident belief that if one emperor was told that a particular art object was the coveted possession of a rival, he could be relied upon to wish to commission

---

[26] Following the assassination of Suleiman's pro-European vizier, Ibrahim Pasha, the Islamic injunction on figural representation was fully complied with. By the mid seventeenth century a French traveller reported that tapestries, European books, maps and globes were stored in the inner recesses of the Topkapi Palace, relics of the past. See Gulru Neçipoglu, 'Suleiman the Magnificent and the representation of power in the context of Ottoman–Hapsburg–Papal rivalry', *Art Bulletin*, **71** (1989), 401–27.

[27] On the documentary evidence for this trip see Spinosa *et al.*, 'To the Glory of Charles V', 81; L. Casali *et al.*, *Gli arezzi della battaglia di Pavia*. See also Necipoglu, 'Suleiman the Magnificent', 419–21; Jardine, *Worldly Goods*, 384–6.

[28] Jardine, *Worldly Goods*, 385.

[29] All kinds of art dealers in fact attempted to drum up Ottoman patronage in the period 1523–36. See Necipoglu, 'Suleiman the Magnificent', 420: 'Aretino's play *La cortegiana* (1534) provides evidence concerning the presence in the Ottoman capital around 1532–33 of foreign sculptors and painters, who followed the Venetian goldsmiths Luigi Caorlini and Marco di Niccolò to Istanbul under the protection of Alvise Gritti's generous patronage. This sudden influx of foreign talent in Istanbul appears to have been precipitated by an invitation that Alvise Gritti extended in 1532 to Pietro Aretino and his artistic circle, immediately after the humanist offered his services to Ibrahim Pasha and through him the sultan in a letter dated 2 August 1531. Insistently inviting him to Istanbul, Alvise urged him to bring along as many of his associates as he desired, including friends and servants, in return for lucrative pensions that no other prince could offer.'

an equivalently impressive art-work for himself. What the Constantinople trip also shows us is that this contest for art-works had an easy currency at the point of encounter with supposedly utterly hostile, alien imperial powers to the east. It shifts the cultural centre of gravity of such activities firmly toward the oriental. We now see that not only the virtual aesthetic space (figured in the tapestries), but the actual space of aesthetic encounter in the era, lies along an axis of commerce and competition from mainland Europe to the eastern seaboard of the Mediterranean. However antagonistic and aggressive their positions on international politics and religious doctrine, imperial rulers became mutually respectful colleagues, with shared understanding and interests, within the international marketplace for top-of-the-range luxury goods.[30]

The imperial gaze: the 'Conquest of Tunis' tapestries

Probably the most vivid realisation of this assumption of a shared (or at the very least recognised) space of aesthetic power-struggle which breaches the boundaries between West and East is a series of tapestries commissioned by Emperor Charles V on his own behalf, and celebrating his own fame – although as usual, his bankers put up the money, and the purchase was eventually added to the imperial 'tab' of permanent indebtedness. This series, too, depicted a momentous imperial Habsburg victory, this time over the Turks at Tunis in 1535. On this occasion, however, one may reasonably state that the entire event was staged as a propaganda coup, and the tapestries designed to carry the image of Charles as 'scourge of the infidel' throughout the known world.

I referred earlier to Charles having such confidence in victory at Tunis that he brought along artists to witness and record it. One of these men was van Aelst, recently returned from Constantinople. The other was the artist Jan Cornelisz Vermeyen, from Haarlem in the Netherlands. In June 1535, Charles hired van Aelst and Vermeyen for a project not dissimilar to the one in which van Aelst had tried to interest Suleiman the Magnificent two years earlier. The two men's job would be to travel with the Emperor on his North African campaign, to make drawings recording its progress, and to turn these into tapestries celebrating his resounding triumph.[31]

---

[30] See Brotton, *Trading Territories*; Jardine and Brotton, *Global Interests*.
[31] For a fuller version of the story of the 'Conquest of Tunis' tapestries and their manufacture, see Jardine, *Wordly Goods*, 386–96.

FIGURE 5.5 'The Mediterranean' from the 'Conquest of Tunis' map tapestry
sequence by Willem de Pannemaker. © Patrimonio Nacional, Madrid.

The Emperor's fleet of 400 ships, carrying an army of 30 000, sailed from
Cagliari on 14 June, under the leadership of Admiral Andrea Doria. By late
July Tunis had fallen, as anticipated, and Charles had ordered the sacking of the
city – a deliberately brutal punishment for the fact that the town had refused
to surrender. Charles' victory, and his subsequent installation of a puppet ruler,
as vassal subject to the Habsburgs, was a huge propaganda coup, confirming
him, finally, as 'Holy Roman Emperor' in deed as well as in name.[32]

The first tapestry in the 'Conquest of Tunis' sequence provides a resonant
symbol for the impact of the campaign, extending imperial Habsburg dominion, as it did, beyond the frontiers of Europe, southwards toward Africa (see
Fig. 5.5). The tapestry depicts a defamiliarised map, which takes as its vantage
point the seaboard of the port of Barcelona (which is behind us, out of sight).
From here the Habsburg army set out for Tunis. Viewers look toward the theatre of military operations, with Spain behind them, as the inscription on the
tapestry carefully explains:

---

[32] Charles had been 'elected' Holy Roman Emperor only by virtue of enormous cash
settlements with the individual electors, for which Charles remained indebted to the
Fugger bankers throughout his reign.

In the distance the coasts of Africa, like those of Europe and its boundaries, are seen with their chief ports, their broad gulfs, their islands, their winds, at exactly the same distances which they really lie, the author having taken more care over their precise situation than over the requirements of painting. All has been done, for the countries as well, in strict accordance with cartography/cosmography, and the painter has observed the canons of his art, considering that the spectator's viewpoint is from Barcelona, where the embarkation for Tunis took place. This town lies between the spectator and the south, with the north behind, over the right shoulder.[33]

Facing toward Africa, the Emperor places the familiar topography of Europe behind him, and looks across the water toward the unconquered Turkish territories to which he lays claim – anticipating becoming 'Lord of all that he surveys'.[34]

Jerry Brotton and I have told the story of the subsequent design and manufacture of the 'Conquest of Tunis' tapestry series at length elsewhere – remarkably detailed documentary sources survive telling us how much the material cost, who paid, who lent to cover the payments, how long production took, and who finally took delivery. Conceived of from the outset as a lasting monument – as propaganda – there was endless wrangling over design detail and quality

---

[33] This information is contained in the cartouche to the right of this extraordinary image, held by the figure of the cartographer, on which the full inscription runs: 'The conquest of Africa in 1535 of Charles the Fifth Holy Roman Emperor and the first King of Spain of that name . . . As it is necessary for a clear understanding to know the country in which the events took place and what preparations had been made, the action is treated in this tapestry according to nature, all that concerns cartography/cosmography, leaving nothing to be desired. In the distance the coasts of Africa, like those of Europe and its boundaries, are seen with their chief ports, their broad gulfs, their islands, their winds, at exactly the same distances which they really lie, the author having taken more care over their precise situation than over the requirements of painting. All has been done, for the countries as well, in strict accordance with cartography/cosmography, and the painter has observed the canons of his art, considering that the spectator's viewpoint is from Barcelona, where the embarkation for Tunis took place. This town lies between the spectator and the south, with the north behind, over the right shoulder. With accuracy thus established, the peculiarities of the other tapestries can be better understood.' See Brotton and Jardine, *Global Interests*, 82–113.

[34] This phrase deliberately evokes Marlowe's play *Doctor Faustus* – a play which, at the very end of the sixteenth century, recapitulates with particular vividness Charles V's global ambitions – for example: 'Faustus, these books, thy wit and our experience / Shall make all nations to canonize us. / As Indian Moors obey their Spanish lords, / So shall the spirits of every element / Be always serviceable to us three' (I.ii.119–23). Like Satan tempting Jesus in the wilderness, Mephostophilis takes Faustus to the top of the highest mountain and promises to make him Master of all the lands he sees spread out beneath him.

control; and the final contract for weaving the series, drawn up between Mary of Hungary and a leading Brussels tapestry maker, Willem de Pannemaker, was not signed until 1548.

By the time the 'Conquest of Tunis' tapestries were completed, the victory itself was nothing more than a minor milestone along the route of the imperial fortunes of Charles V. As art objects, however, the tapestries were sensational masterpieces, whose lavishness and beauty of execution made them appropriate monuments to the enduring magnificence of the Habsburgs. Huge, but conveniently portable, they travelled with the Habsburg retinue (which was constantly on the move, because of the sheer size of the Habsburg dominions to be administered), so that those who visited the imperial quarters could be dazzled by the splendour of the tapestries, whilst at the same time registering, panel by panel, and depicted with awesome visual realism, the Emperor's formidable might.

Pannemaker completed the 'Conquest of Tunis' series in April 1554, nearly twenty years after the campaign it so vividly and meticulously depicted. In the 'Conquest of Tunis' tapestries, art and empire fuse. The 'Conquest of Tunis' stands (as it had been commissioned to do) for a key moment in the political struggles of the Holy Roman Emperor with the infidel, as well as for the sheer, breathtaking, political, economic and cultural power and influence of the German–Spanish imperial line.

In the summer of 1554 the tapestries were packed up in Brussels under the supervision of Simon de Parenty, and their manufacturer Pannemaker accompanied them to London, where they arrived on 3 July 1554. The entire series made its first public appearance at the wedding of Charles' son Philip II of Spain to Mary Tudor, which took place in Winchester Cathedral on 25 July 1554.[35]

In England, Philip was a deeply unpopular choice as consort for the Queen regnant. The marriage was widely seen as another act of Habsburg aggression, a bid to annex England to their Europe-wide Catholic Empire. And indeed, astutely brokered marriages had for several generations been one of the Habsburgs' most effective means of consolidating their territorial holdings.

[35] See Alex Sampson, *The Marriage of Philip of Habsburg and Mary Tudor and Anti-Spanish Sentiment in England: Political Economies and Culture, 1553–1557*, unpublished Ph.D. thesis, University of London (1999).

Philip's first marriage to his first cousin, Maria of Portugal, had, as we have already seen, secured him rights to that country (which would, indeed, later be translated into his actual claim to sovereignty over Portugal in 1580); no less was to be expected from Philip's marriage to the English Queen regnant.

Nor were matters helped by Charles V's publicly stated vow to stamp out the 'heresy' of Protestantism in Europe, just as, symbolically, in the victory at Tunis, he had humbled and laid low the Ottoman 'infidel' to the east. Charles' personal ambition to establish the world-wide reach of Christendom in the name of a restored Holy Roman Church did not augur at all well for England, a country which had lately had two Protestant kings, and which now saw itself about to be ruled by the Catholic Charles' firstborn son.[36]

The twelve panels of the 'Conquest of Tunis' tapestry series narrate, episode by episode, the symbolic enactment of Habsburg power, and the Habsburg realisation of a long-promised scourging of the turbaned, infidel Turk. Proudly displayed for the first time for the Anglo-Habsburg marriage, the 'Conquest of Tunis' tapestries can have done nothing to allay English fears concerning the implications of the royal union. Substitute Protestant 'infidels' for Turks – as European Protestants were increasingly inclined to do – and its message must have been clearly read by those admitted into the presence of the royal newly weds. The Habsburg–Tudor couple would – and did – scourge the 'infidel' Protestant with all the ferocity directed at oriental adversaries.[37]

An epilogue on the (movable) Spice Islands

Let me draw my argument to a close. In the Renaissance tapestries we have been considering, art is inseparable from power; the aesthetic space maps deftly onto the territorial and political claims of the rival imperial powers. The competitive acquisitiveness embedded in the whole idea of Renaissance art

---

[36] As mentioned above, in 1580, long after the death of Mary Tudor, Philip II did indeed become King of Portugal, on the strength of his first marriage to Maria of Portugal. Thus English anxiety was well grounded. Had the union with Mary Tudor resulted in children, a Habsburg claim to England would have been a real possibility, in spite of attempts in the drafting of the marriage contract to rule out such a claim.

[37] On the easy, habitual substitution of Protestant for Muslim in the religious polemics of this period see Sampson, *The Marriage of Philip of Habsburg and Mary Tudor*.

FIGURE 5.6 Globe detail from third 'Spheres' tapestry; right, Diogo Ribeiro planisphere for Charles V, 1529. © Patrimonio Nacional, Madrid.

objects – exchanged, transacted and desired as objects of beauty and status across ideological and doctrinal boundaries – means that their centre of operations has to be thought of as located well to the east of where we are accustomed to see it. On the space mapped by Renaissance cartographers, on maps and globes whose value corresponds to the grandeur of their claims, the centre lies

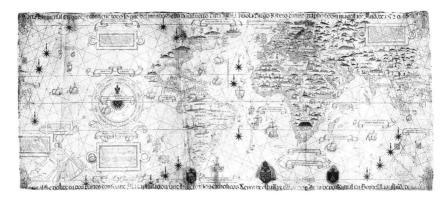

FIGURE 5.7 Diogo Ribeiro planisphere for Charles V, 1529. © Biblioteca Apostolica Vaticana (Vatican).

on the eastern seaboard of the Mediterranean, where Habsburg and Ottoman dynastic aspirations met.

I must emphasise that the mapped spaces of the tapestries were not merely virtual, aspirational. Let us turn one last time to the globe figured in the third 'Spheres' tapestry.

At the extreme right of this map are the Moluccas – the 'Spice Islands' (see Fig. 5.6). As we have seen, although the 'Spheres' tapestries commemorated a wedding which took place in 1525, their production took some number of years thereafter. Their manufacture therefore coincided with a period of intense competition between Portuguese and Habsburg cartographers concerning the precise location of the Moluccas on the map. At issue was on which side of the notional line of demarcation between agreed Portuguese and Spanish Habsburg dominion – the Tordesillas line – the islands were located.[38]

In 1529 Charles V and his brother-in-law John III of Portugal sat down at the negotiating table in an attempt to resolve the vexed issue of possession

---

[38] 'The Treaty of Tordesillas, which had been signed in 1494 in the aftermath of Columbus' discoveries in the western Atlantic, had provisionally settled the territorial dispute between Portugal and Castile as to where their mutual spheres of territorial influence extended. However, by fixing the line of demarcation 370 leagues west of the Cape Verde Islands, in the Atlantic, it only postponed the inevitable diplomatic dispute as to where the line should fall in the eastern hemisphere, if it was to be extended and drawn right around the globe. The treaty was clearly drawn on a flat map of the known world, rather than a globe, which would have allowed the two crowns to partition the world like splitting an apple in two' (Brotton, *Trading Territories*, 120).

of these commercially priceless islands.[39] On the table was a new map, produced by the brilliant Portuguese cartographer Diogo Ribeiro, which had,
in the light of Magellan's circumnavigation of the globe in 1522 (on which
Ribeiro had served as chartmaker[40]) relocated the Moluccas so that they lay
within the half of the globe defined as belonging to Charles V (see Fig. 5.7).[41]
Hitherto, the islands had been drawn on the Portuguese side of the dividing-
line, and negotiations had been based on a Portuguese claim to ownership.
Now Charles graciously 'yielded' the Moluccas to the Portuguese for a substantial sum (300 000 to 350 000 crowns, it was rumoured) in hard cash. The
wording of the treaty drawn up on 17 April 1529 emphasised the part the
redrawing of the map had played in establishing diplomatic and economic
concord:

> In order to ascertain where the said [Tordesillas] line should be drawn, two
> charts of the same tenor will be made, conformable to the chart in the India
> House of Trade at Seville, and by which the fleets, vassals, and subjects of the
> said Emperor and King of Castile [Charles V] navigate. Within thirty days from
> the date of this contract two persons shall be appointed by each side to
> examine the aforesaid chart and make the two copies aforesaid conformable
> to it ... This chart shall also designate the spot in which the said vassals
> of the said Emperor and King of Castile shall situate and locate Molucca,
> which during the time of this contract shall be regarded as situated in such
> place.[42]

Produced under the patronage and with the financial support of a great
imperial figure, Renaissance art objects shaped that culturally specific mapping
of space which we look back upon today – that cultural 'rebirth' and 'flowering'

---

[39] For the full story of the 1529 settlement, and the navigational run-up to it, see Brotton,
Trading Territories, ch. 4, 'Cunning cosmographers: mapping the Moluccas', 119–50.

[40] Unlike Magellan, who died en route, Ribeiro actually had completed the round-the-
world journey.

[41] 'Ribeiro's map of 1529 reflected the increasingly fine distinctions which were by this
stage separating the two crowns ... Ribeiro proceeded to place the Moluccas just
within the Castilian sphere, a striking example of the extent to which he was aware of
how sensitive was his placement of the islands. Ribeiro positioned them 172 degrees
30 minutes W of the Tordesillas line of demarcation, just 7 1/2 degrees inside the
Castilian sphere ... Ribeiro had skilfully managed to produce a map in line with
the rapidly changing requirements of the Castilian claim to the Moluccas' (Brotton,
Trading Territories, 142–4).

[42] Quoted in Brotton, *Trading Territories*, p. 137.

which produced our modern European aesthetic. Yet the motivation behind their making was far from a simply aesthetic one. Like Diogo Ribeiro's tactful redrawing of plum territories so that they fell within the sphere of influence of his imperial patron, tapestry representations opportunistically adapted to the political needs of their owners. Ribeiro's opportunistic act figures particularly vividly the lasting power and influence of Renaissance mappings of space.

6 International space

NEAL ASCHERSON

> It is not what they built. It is what they knocked down.
> It is not the houses. It is the spaces between the houses.
> It is not the streets that exist. It is the streets that no longer exist.
>
> James Fenton, 'A German requiem'

The topic of international space is like one of those monstrous catfish which used to loaf around the hot-water outfalls of the Berlin power stations. You could hook it, net it, spear it, or even seize it in your arms if it were not so heavy and slippery. In an effort to get some grip on international space I have followed Fenton by asking, in three different ways, whether it might be between things – not the houses but the spaces between the houses. It could be looked at as a system of gaps, blanks or crevices between social or diplomatic entities. Or, in an Einsteinian space-time way, as the space which appears when a social or diplomatic house or houses fall down: a vacancy measurable in terms of power, geography and time – usually a pretty short time. Or, lastly, it is possible to understand it as a sort of air pocket, like the spaces which preserved a few fortunate people in the recent Gujarat earthquake. What interests me is the space which opens out or is excavated under the reinforced concrete of a tyranny, and within which human behaviour can regain authenticity.

The international community, as we are still rightly reluctant to call it, is a cellular structure whose most obvious spatial component is the nation-state. Most people would probably prefer to hear about concentric multiple identities than about nation-states, but I am discussing international space, not subjective affiliations. Certainly, there is concentricity to be found in international space.

Space: In Science, Art and Society, edited by F. Penz, G. Radick and R. Howell.
Published by Cambridge University Press. © Darwin College 2004.

Some states contain sharply defined political regions or are formal federations, while many – especially in Europe – are engaged on a programme of pooling sovereignty under some supranational council or commission. This is only part of the historic weakening of the nation-state, as its once absolute authority leaks downwards to the regional level and upwards to the transnational plane. But it is important to recognise that this weakening, which will eventually transform the texture of international space out of recognition, is only in its early stages. For the moment, the political world we inhabit is a cellular honeycomb of nation-states, though some cells are far bigger than others. In power terms, the cellular image can dangerously misrepresent the substance of international relations, in that it suggests a sort of fictional one-state one-voice equality, on the UN General Assembly or OSCE model: it does not illustrate the forces of penetration and control which one strong imperialist state or group of states can exert over others. But international space and its divisions have to be understood in the first place as a form.

Cells have walls. International space, in recent times, has presented a pattern of territorial boundaries, some artificially delimited and others – like the English Channel or the Drakensberg mountains – so-called 'natural boundaries'. The important question is how permeable these cell walls are. At least, it has seemed the important question to Europeans and those under their influence for the last 300 years or so, and to the post-colonial continents for the last century. This was plainly not always so. The notion of physical boundaries disappears into the past, but most of them were relative rather than total. Any pastoral community knows precisely the stone outcrop and the thorn bush where one common grazing ends and some other neighbouring pasture begins, but the rules about where goats can go in such societies are not necessarily the same as the rules about where people can go. Again, Hadrian's Wall was not the Berlin Wall – a total partition – but something more specialised, at once a motorway toll-station and a radar early-warning line made of stone, timber and earth. Roman emperors constructed similar barriers elsewhere, for example in Germany or the Danube Delta. But apart from obstacles like the Danube, the Alps or the High Caucasus, ancient empires were much less clearly demarcated than the close-packed small farms of Attica or Central Italy in classical times.

It would be a waste of time to narrate the hardening of the cell walls, the advent of the idea of physical frontier barriers which were more than just tollgates. That is part of the story of the early modern nation-state, a Eurocentric story whose most interesting and formative event was the 1648 Peace of Westphalia which ended the Thirty Years' War, laying down the principle of non-intervention in the internal affairs of sovereign states. (There is a case for saying that Europeans lived in the Westphalian system until the 'humanitarian intervention' in Kosovo in 1999.) But it is interesting to watch the parallel development of what could be called a 'cellular discourse' outside the strictly political field, especially in the later nineteenth century.

Rudolf Virchow (1821–1902) is mostly remembered for his proposition that human tissue is composed of cells. His most celebrated saying is that 'omnis cellula ex cellula' – every cell is derived from another cell. (Actually François Raspail said it first, but Virchow made it much more famous. On the other hand, Virchow has more streets named after him but none as intellectually majestic as the Boulevard Raspail in Paris.) What is less remembered about Virchow is that he was a political thinker – in Prussian nineteenth-century terms, a liberal. He thought of the human body as a republic of cells, sovereign and equal, and he fought for his republican ideas during the Revolution of 1848, the 'Springtime of Nations'. Later he was elected to the Prussian Diet and then the Reichstag. Bismarck, who thought that international space was a pond full of luscious carp rather than a noble republic of cells, loathed Virchow and tried unsuccessfully to make him fight a duel.

Virchow was also an archaeologist, or more accurately, he was one of those fine old prehistorians for whom physical anthropology, ethnology and archaeological research could still form a single subject. The first true German archaeologist, who insisted that archaeology (*Urgeschichte*) should be an exclusive discipline out on its own, was Gustaf Kossinna (1858–1931). Kossinna despised Virchow's vision of a world-wide 'prehistoric anthropology'. Instead, his interests focused on the material culture of the 'Germanic' past. He was much younger than Virchow, and – like Bismarck and for some of the same reasons – he detested him. Kossinna, a conservative right-wing nationalist, had no patience with the school of historians who wanted to fit Germany's early past into the context of classical Greece and Rome, or with the *Römisch-Germanisch* archaeologists, based mostly in the Catholic Rhineland, for whom the only respectable aspect of German early history was the contact with the Roman Empire. A fervent racialist, Kossinna disliked both Virchow's holistic approach to learning and his internationalism.

Kossinna, all the same, was another cellular man. The difference between him and Virchow was in a sense over Virchow's 'omnis cellula ex cellula'. For Gustaf Kossinna, the human cultures which mattered were not interpenetrating but autochthonous, self-creating. They did not derive from the impact of, or mixture with, other cultures (invaders, migrants, intermarriage across ethnic boundaries), but arose and developed as the consequences of innate biological forces in their own 'blood' or gene-pool. A healthy cultural 'cellula' had more or less impermeable walls and its contents were more or less homogeneous.

Christopher Hann, discussing what he called the 'Malinowski period' of early field anthropology, in his book *Social Anthropology* (one of the volumes in the Teach Yourself series), underlined the emphasis laid by Bronislaw Malinowski and his disciples and students on the uniqueness and distinctiveness of each society they inspected – especially small island societies in the western Pacific. 'The world', Hann writes, 'was a mosaic of these bounded "peoples", "cultures" and "societies". This fitted with the idea of the nation, as it was then gaining strength in Europe.'

For 'mosaic', one could substitute the word 'cloisonné' – a technique in which each garnet in (say) the Sutton Hoo belt-buckle is enclosed in a tiny wall of gold. Kossinna, in his own version of the mosaic of bounded peoples, worked out what is still known as the culture-historical approach – his own term was *Siedlungsarchäologische Methode*, 'the settlement-archaeology method', which is much less explicit. In 1911, he summed up his method in a celebrated sentence: 'Sharply-defined archaeological culture areas correspond at all times to the areas of particular peoples or tribes.' In other words, a specific array of tools, ceramic types, hut-building techniques and so on revealed the presence of a specific ethnic group defined not only by its material culture but by inherited racial traits and by language – wherever that array turned up. Find a particular type of iron spearhead, a particular decoration on pottery and a particular way of setting timbers to support a farmstead, and you could exclaim: 'This settlement was Germanic! These people spoke proto-Germanic. Here dwelt our ancestors!'

Kossinna's admirers for a time included the greatest of Marxist archaeologists, V. Gordon Childe. But Childe did not follow Kossinna's most fervent German disciples as they partitioned Eurasia into 'Nordic Culture-Regions' and 'Germanic / Celtic / North Illyrian Settlement Areas'. Nazi archaeology was almost entirely constructed around Kossinna's ideas. But the 'culture-historical approach' also became rapidly and lastingly popular throughout non-German central Europe, where it served (especially in pre-1939 Poland) as a method of demonstrating the antiquity of Slav or 'proto-Slavic' settlement in the territories of the Polish state. The approach was easily adapted by Stalinist archaeology during the Cold War to 'prove' that a wider Slav world had existed in east-central Europe long before it was disrupted by imperialist Germanic invaders. Though generally discredited in the West, strong traces of Kossinna's approach still survive in central and eastern Europe. And the truth is that the world of Anglo-Saxon scholarship still owes more to Kossinna than it likes to admit. When 'progressive' archaeologists in Australia or North America allow their dating results to be used to validate indigenous land claims, almost all of them based on priority of settlement and on ethnic continuity with prehistoric settlers, the ghost of Gustaf Kossinna still rides the night.

Out in the political world, Kossinna's vision of self-creating, hard-walled, homogeneous cells could serve as a guide to the 'Modern Nationalism'. This ideology became hugely influential in the later nineteenth century, though it had long roots into intellectual history. At its core was the view that no state could be strong or secure its rightful place in the sun unless its population was uniform – which usually meant ethnic or imagined racial uniformity. Minorities were a state's misfortune. The Modern Nationalism was especially popular among the intellectuals of countries which were either struggling for independence or in danger of losing it; it expressed a violent reaction against the apparent failure of loose, multinational and multicultural polities to survive in the shark-pool of international conflict.

One good example of this thinking was provided by Roman Dmowski, founder of the Polish nationalist movement called the Endecja – or National Democrats. The old Polish–Lithuanian Commonwealth, which had been abolished by the Partitions in the late eighteenth century, had been just such a plural polity; all that had been required of its citizens, Poles, Balts, Jews, Germans, Ruthenians, Ukrainians, Muslim Tatars or Calvinist Scots, was loyalty to the crown. In the insurrections after the Partitions, men and women from all those communities had fought to regain Poland's independence. But Dmowski argued that the Commonwealth had been an anachronism, an unhealthy nation-cell infected by substances from other cells. In the future Poland, the 'true Pole' could only be a Catholic Slav. Above all, Dmowski preached that the Jews were disloyal, an alien threat to any revived Poland. Dmowski wanted that Poland to be ethnically homogeneous, without minorities – the Poland which in fact emerged after 1945, following the genocide of the Jews, the annexation by the USSR of Belorussian and Ukrainian territories and the expulsion of the Germans.

Cellular, too, was the vision of the Kemalists in Turkey, who eventually constructed a supposedly mono-ethnic state out of the rump of the vast Ottoman Empire, with all its plethora of semi-autonomous minorities, faiths and languages. This was Dmowski's ideal in action, the first time that the fashionable Modern Nationalism was put into practice. The Armenians were slaughtered, the Greeks expelled, the Kurds suppressed; the racialist fiction of Turkey for the Turks was imposed. And the notion of ethnic homogeneity as the precondition for a strong nation-state – so consequently practised by Hitler and Mussolini – still dominated the outlook of democratic statesmen at the end of the Second World War. Churchill felt guilty about the Allied betrayal of Poland; he tried to make amends by helping to clear some 6 million Germans out of Silesia and East Prussia so that the new Poland could at least start with the benefit of a united, 'Polish' population.

But cellular nationalism of the Kossinna/Dmowski kind, although disastrously transmitted to many continents by colonial empires, is on the way out. That does not mean that globalisation will reduce international space to a single space. On the contrary, all the evidence is that global markets and communications actually set off a proliferation of cells. The UN had 51 founder members in 1945; by the end of the century it had 192, of which 46 (counting the Vatican) were microstates. Nearly a quarter of the world's nation-states have fewer than a million inhabitants, some – like Palau with 15 000 – far fewer. This reorganisation of international space is not a simple 'reaction against bigness'. It has, more importantly, come about because a globalised world economy, free of all but local wars, is a perfect environment for microstates as long as their cell walls are porous. This fragmentation is related to the larger process of the decay of the traditional nation-state, as its authority leaks away upwards to the supranational level and downwards to the more adaptable, sustainable level of regions.

The honeycomb, however, is not the only image for international space. There are several alternative ways of defining it. To start with, try looking not at the cells, but at the space between the cells – not the houses, but the space between houses. On the atlas, there is not much white space left. Antarctica remains a sort of common territory, on paper the site of all kinds of ambitious sectoral claims by nation-states but in practice a place of reasonably friendly, treaty-bound co-operation between research stations. The main body of the oceans, their floors and their habitats, remain untenanted international space which will soon – not without struggles – have to be put under some much more restrictive international regime.

In the past, in the early twentieth century, there were a number of spaces which were not absolutely unpopulated but whose allocation to empires or nation-states was undecided. Some of them produced postage stamps of strange shapes, decorated with obscure overprints. There were Anglo-French condominia, and Tannu-Tuva, and that diamond-shaped space drawn on the map of Mesopotamia within which RAF biplanes bombed nomads. The Aland islands in the Baltic became one of the few complete successes of modern diplomacy, as Finnish sovereignty compromised with the virtual independence of the Swedish-speaking population. Later in the century, under the increasing threat of renewed European war, less durable spaces appeared between the national cells, like the so-called Free City of Danzig with its League of Nations High Commissioner, its German population and its Polish post office. (Odd to recall that the judge who tried Joschka Fischer, now German foreign minister, for rioting in 1970s Frankfurt was the judge who condemned the defenders of that post office to death in 1939.) International space was pulled into some desperate shapes: the Caprivi Strip in southern Africa, or the Polish corridor to the sea, resembling the siphon of some bivalve – a bit like the little Moldova corridor which has been drawn to give the new state 500 metres of Danube bank and 'access to the oceans'.

But the most intriguing intercellular spaces are just gaps, crevices, interstices, oversights. They appear whenever some new international system attempts to demarcate everything sharply, menacingly and in a hurry. For example, there may or may not have been something called the 'Akwizgran Discrepancy'. A forgotten thread of diplomatic folklore suggested that when the new Kingdom of Belgium emerged in 1831 – much to the annoyance of the Congress Powers who had imposed the Vienna settlement on Europe after 1815 – there had been a demarcation error at the point where the borders of Belgium, Germany and the Netherlands met. Somewhere between Aachen and Verviers, there existed a tiny triangular space, big enough to contain a house, a patch of field and a few fruit trees, which belonged to nobody. During the Cold War, the Polish writer Stanislaw Dygat got past the censor with a romantic novel about love and freedom in the Discrepancy – if only! He did not have to spell out his point. I have never been able to find the Discrepancy, which probably never existed. But the thought of it was dear to people.

In the same way, the very absoluteness of the Cold War borders left a whole series of chinks and crevices. Some of the best were in Berlin, naturally. There was a whole archipelago of tiny islets of West Berlin territory floating offshore in East Germany, connected to their mainland by barbed-wire causeways. There were also places in what had been the city centre where the Wall had been pulled back a little to make it more defensible, leaving patches of East Berlin territory on the Western side. One of these patches was the ruin of the old Potsdam Station, a sinister precinct inhabited by spies, petty crooks, hermits and dope-dealers. Sometimes they lit fires at night. Few Westerners ventured in there. Legend warned that *Volksarmee* patrols occasionally dropped over the Wall and seized anyone they found in the ruin as an 'illegal immigrant to the German Democratic Republic'.

Not far away was the Lenné Triangle. Only the pavement kerbs marked where buildings had once stood; it became a fenced-off wedge of scrubland and small trees. But one day in 1987, the Triangle was occupied by a tribe of 'alternative' people ('autonomes') in flight from the West Berlin police, who set up a shack and tent city among the bushes, parked their Deux Chevaux against the Wall and invited everyone to come and join them in no-man's-land, the space of freedom between worlds. Visitors poured in, to smoke dope and plan new ways of living or merely to watch autonomes making love in the 'Biotope'. It was too good to last. But when the West Berlin police finally stormed the encampment, East German soldiers suddenly appeared along the top of the Wall and helped the occupiers to escape.

Close to international 'space between' is international 'space after': the space or gap which appears when something is suddenly removed – Einsteinian space-time. 'It is not the streets that exist. It is the streets that no longer exist.' A front tooth is pulled, a mosque in the main square is dynamited to rubble, a multinational kingdom collapses. The difference is that when we are talking about a 'space after' suddenly appearing in a national or international order, rather than in a human mouth or a Bosnian townscape, it has aspects of a vacuum rather than a mere vacancy. The space creates a limited time, as competing forces rush in to fill the vacuum and replace it with new or reinvented structures. When four empires fell apart or were destroyed within a few years, between 1917 and 1919, an inrush began to fill the enormous emptiness they left behind, the inrush of what came to be called 'successor states'. But it is not just the removal of a state which produces this suction. Assassination can have the same effect. When the Basque ETA car bombers killed Franco's designated successor, Admiral Carrero Blanco, the possibility of any orderly transmission of fascist power to another ruler before the aged Franco died was instantly removed, and the inrush to fill the space became a transition to democracy.

Space, here, is being used as a metaphor. Obviously, the sort of international space created by the removal of something, or by the appearance of an overwhelming opportunity, is not literally empty or uninhabited. It is not a space to the indigenous people who live there, whether they are Bushmen, Aboriginals or rural Belorussian-speaking villagers who answer questions about identity by saying: 'We are *tutejszy* – we are from-here people.' Sometimes it is the indigenes themselves who try to establish mastery over their own place when an external authority has fallen away, as the Poles and Czechs did after 1918, or as the Armenians tried and failed to do in the disintegrating Ottoman Empire in 1915. More often, in modern history, outsiders have entered and declared: 'This is terra nullius, no-man's-land, and this handful of savages wandering across its surface will become our colonial subjects.'

The colonising discourse is full of exaltations about 'emptiness' and 'wide open space'. But the Canadian North, the American West, the Cape or the Australian outback, all celebrated by Europeans as 'virgin' and 'empty' space, were the traditional territories of human beings who had been there for millennia. Most of those first people were hunter–gatherers. Those who were agriculturalists or nomadic pastoralists had already encountered the colonists as they established themselves near shores and estuaries, before later settler generations embarked on the big trek into the grasslands and temperate forests of the interior.

The terms 'open space' and 'emptiness' served, of course, as the justification for the global land grab which followed, often accompanied by genocidal massacre. It was as if the notion of occupancy, let alone of collective ownership of land, could not apply to hunter–gatherer communities, who to this day have been denied the right to full title over indigenous land which is basic to farming and ranching societies. This use of the concept of space reduces a whole category of humans to sub-humanity. It was one thing for Afrikaner politicians to maintain that the Bantu peoples were newcomers who had entered an 'empty' South Africa from the north only a few years before the entry of Dutch settlers from the Cape. That was simply a lie. It was much worse to treat traditional hunting territories used by the Bushmen at the Cape as if they were unclaimed property, on the grounds that people who lived as they did were too primitive to have title over anything beyond personal property. In *The Other Side of Eden*, the anthropologist Hugh Brody postulates millennia of conflict between hunter–gatherers and agriculturalists. It is the farmers, he asserts, not the hunters, who are continuously on the move. The farmers are the true 'nomads' as they push compulsively onwards and outwards in search of fresh soils, expelling the native hunter–gatherers to the miserable deserts or tundras beyond the margins of the cultivable world. The farmers change the environment; but the hunter–gatherer is part of it. (This view of the coming of agriculture as a result of migration is entirely rejected by 'processual' archaeologists. But as an account of how white colonist farmers displaced indigenous peoples, it works.)

The phrase 'time out of mind' was used in England mostly about space. It was a legal phrase about customary leases, which may even have survived until the abolition of copyhold tenure in the twentieth century. But it was an example of how the notion of time is plaited into the notion of assigned space. The danger that assigned space could suddenly turn into a gap, or vacuum, sucking in undesirable new occupants was always obvious. Accordingly, especially in settled societies, there have usually been ways in which time is used to plug a sudden gap in space.

A tenant dies, but his heirs appeal to 'time out of mind' for the right to inherit the tenancy at the same peppercorn rent. The Pequod nation is reduced to a handful of people living on a scrap of reservation in the Connecticut woods. But one old Pequod lady proves capable of invoking time – documented continuity in a homeland since the seventeenth century – to support the fabulously successful series of land restitution claims which created the large, wealthy and autonomous space at Mashantucket where the Pequod nation flourishes today.

Reporting from Poland, I have seen many examples of time being used to patch over a dangerous breach in space. For most of my life, the Royal Castle at Warsaw was no more than a fang of brickwork left by the Nazi dynamiters in a monstrous urban gap. But in the 1970s it was rebuilt down to the last detail, and the forty years when the Castle did not fill its space are dismissed almost as an optical disturbance. 'Look across this courtyard', says the guide, 'and you can see the only Renaissance window which survived the Baroque reconstruction'. At Gdansk, during martial law, I watched newly married couples laying their bridal flowers under the Solidarity monument, erected the year before to commemorate martyred shipyard workers. I said to my Polish companion that this was a freshly invented tradition. He glanced at me, puzzled, and then replied: 'Well, but in a way the monument has always been there.'

Episodes like those remind me of the late Edmund Leach's work on the topology of elastic forms. Time is an elastic sheet which can be used to pull the edges of wounds together. I remember the elderly man I met leaving a Madrid polling station in 1976, during the first free elections in Spain for over thirty years. I asked him how he had voted. 'I voted Socialist', he replied, 'I always vote Socialist.'

Space can also imply a sufficiency of territory. International space has often been a phrase in disputes about how much space is enough. The tide has gone out on most of those arguments, leaving a few weedy posts in the mud. It used to be said – reassuringly, as I recall – that there was just enough space on the Isle of Wight for the entire population of the world to stand shoulder to shoulder. There is no longer enough space. In any case, there is nothing reassuring now about a human race with standing room only. 'Enoughness', the idea of a sufficiency of space, has too many applications and variables to be useful. Enough in stony uplands is much bigger than enough on river-bottom alluvial land. Preparing for war in 1914, the German Empire aimed to absorb geographically tiny territories from France, the sites of strategic factories or minerals, but colossal regions from the Russian Empire. The suggestion that a nation requires more space has always been the prelude not only to crime but to failure; partly because the suggestion at once unites a coalition of hostile neighbours, but even more because the obvious question – 'Space for what?' – practically never gets a coherent answer. The German eastward colonisation across the Elbe in the early Middle Ages did succeed, because it was nothing like a planned Lebensraum programme: it was driven by the separate ambitions for land and trade of a large number of individuals and corporations. In contrast, the planned colonisations undertaken by Bismarck and Hitler left shelves of theoretical works about Raum but little else save graveyards. The Poles in West Prussia and Posen bought back their own land from the German colonists, who thankfully streamed home again (*Ostflucht*). The population transfers and ethnic plantations attempted by Hitler, in the name of exploiting the huge new space cleared by conquest, also came to nothing. That space found only one effective use; its relative remoteness made it a suitable site for genocide.

And yet the longing for enough space, family longings rather than national longings, changed the world in the nineteenth and twentieth centuries, and is still doing so. The 'huddled masses' began to move, heading above all for North America: Jews from congested shtetls, Gaelic Highlanders escaping land hunger and rack renting to own a rectangle of uncleared forest in Ontario, and the post-Famines outrush from Ireland. Those European generations did, for once, have an idea of what sufficiency of space might mean. The 160 acres offered to settlers by the 1862 Homestead Act in the United States became something like a universal standard.

These people had set out with a vivid awareness of the physical and moral stunting inflicted on their own parents and children by lack of space. It was the writer Artur Sandauer who used to tell of a superstition among Volhynian Jews. They believed that at the moment of birth an angel appears by the baby for a split second, and opens its vast wings. If the wings can be fully spread, the baby will enjoy a free and happy life. But if the room is too small, then the child's life will be narrow, poor and frustrated. The wise men therefore reasoned that it was best to be born in the open air. Sandauer, who counted himself happy, remarked that he had been born when his pregnant mother had been fleeing from the Tsarist police toward the Austrian frontier. Her sledge turned over, rolling her downhill into a snowdrift where she went into labour. I liked this story, and repeated it to the first person I met, who happened to be a Krakow taxi-driver. He nodded without smiling, and said: 'I was born when my mother was in a Gestapo cell.'

The mother who fell off the sledge was in transit between spaces – from Romanov space to Habsburg space. They were not identical, which was why she was making the journey, but with difficulty they were compatible and mutually accessible. (There used to be a picnic place near the Upper Silesian city of Gleiwitz, now Gliwice, which was called the *Dreikaisereck* – 'Three Emperors' Corner'. Families took photographs there, posing by the stone, which pinned together the domains of the Hohenzollern Kaiser, Franz-Josef of Austria-Hungary and Tsar Nicholas II.) But there have also been categories of international space which were mutually impenetrable. This was not because of dense rainforest or deserts. It was because different societies could conceive of territorial and cultural space in styles so dissimilar that imagination could not cross between them.

My favourite book on this subject is the strange and brilliant work by François Hartog called *The Mirror of Herodotus*. Hartog discusses the contrast, as observed by Herodotus, between the Scythian pastoral nomads of the Black Sea steppe and the trudging infantry battalions of the Persian Empire under the command of Darius. When Darius invaded Scythia, he was unable to bring the ever-retreating Scythians to battle and was baffled (as were the Greeks) by this encounter with peoples who had no fixed centre, no capital or inner sanctum. He had built a bridge across the Danube as a *poros*, a means of intercommunication between spaces. But he could not get at the Scythians because they were *aporoi* – 'incommunicable'. Their space was an *aporia*, a limitless grassy universe which was trackless, disorienting, 'ungetatable'. Soon the Persians only wished to find their way safely back to their *poros* and across the river again, before the alien space swallowed everyone. Hartog goes on to suggest that this was also true of cultures: Greek/Persian and Scythian cultures were *aporoi* to one another, so that once you had ventured decisively into the other space you could not get back again. Again he quotes Herodotus, who told the fables of Anacharsis and Scyles, two Scythians who in different ways crossed into the space of Greek religion and lifestyle and who both met death when they returned: Scyles handed back to his own people as a hostage and executed the moment he reached Scythian space, Anacharsis making a clandestine landing on the Crimean coast but killed at once by the arrow of a king.

This imagining of a space, which is 'other', in which the man or woman who enters may be irretrievably changed, which by its very nature is irreconcilable with adjoining space, leads me to the idea of space as breathing-space, a survival chamber hollowed out within the foundations of an oppressive system, something like the vacuole within a cell of plant tissue. The Early Christian catacombs were literally such spaces, subterranean galleries where Rome was not present and could not enter physically or metaphysically.

These are spaces of *authenticity*. Within them, whether they are physical or social or spiritual, people escape prevailing constraints and can behave spontaneously, truthfully, in accordance with what they feel to be their real nature. 'Real' is a significant word here. The space outside and the laws which obtain there are judged to be in some way unreal. This may be because that world out there is governed by persons who dwell in darkness; they have not received the enlightenment which reveals the real order of creation and divine authority. Or it may be because the world out there is the sort of tyranny in which everything is available but everything is a fake: fake patriotism and fake history, fake elections and fake public enthusiasm; phoney economic statistics and hypocritical constitutions and imposing, meaningless programmes for scientific research. Even breathing and making love can become as-if activities.

It is at this point that human beings begin to excavate authentic space, in order to survive morally. Often enough, the space is no bigger than a family apartment. German Pietism, for example, has in the past concluded that the best response to satanic regimes is to live a pure and holy life at home, not challenging the public order but bringing up children who do not lie to their parents and who know good from evil. (It was this sort of minimalist resistance, which was not enough, finally, for Dietrich Bonhoeffer during the Third Reich.) And there are even smaller spaces. It was either a Pole or a Czech who said of their Communist systems: 'The one sure way to destroy them is to do your work with absolute honesty and as well as you possibly can.'

A smaller space, but a more ambitious one. Here is the notion that an authentic space is not static – a locked room for the conservation of values – but dynamic: a cave, which can be enlarged until it has hollowed out the very foundations of the ruling order. Authenticity is like a termite or a deathwatch beetle, gnawing away expanding spaces within apparently solid structures. We still do not really understand why the Soviet Empire in Europe collapsed with such brittle abruptness in 1989. But part of the reason is that it had become riddled with authenticity.

These dynamic spaces of authenticity can take many forms. One is conspiracy. Between plotters at first, and then among a far larger network of supporters and organisers and couriers, there is 'real', genuine speech. Not all the truth can be told, but truth is at least the currency. The risk taken is not a fake, and neither is the hope. And when and if the conspiracy breaks surface and becomes an insurrection, then there is often a festival of authentic emotion, of hectic but authentic relationships. The Warsaw Rising of 1944 was like that, a time when masks fell away and the insurgents, reaching out for one another, seemed to discover their own better selves.

The opening of authentic space during the Cold War was less dramatic. Often it required long, patient games to infiltrate authenticity into fake institutions. Party rule meant control of civil society. The local branch of the Slobodnian League of Stamp-Collectors had to submit the candidates for its new committee to 'the appropriate authorities' who would amend the list, inserting extra Party names if necessary, and confirm the presence of one or two individuals who could be relied on as informers.

As time passed, this became harder. Occasionally, an official organisation would turn authentic, or its committee would go native. That is what happened in Czechoslovakia with the Jazz Section of the Czech Musicians' Union. One day, the authorities woke up to find that a Mr Karel Srp and his friends had refunctioned the Jazz Section into a forum where people spoke their minds about politics as well as Brubeck. The section's journal, technically free of censorship as the organ of a state body, became unobtainable as all Prague rushed to read and copy it. It was some time before Mr Srp could be locked up. Much damage was done, but this technique was not confined to Communist Europe. In Franco's Spain, the Communist Party contrived to re-function the Falangist trade unions into spaces of authenticity, by inserting into them the genuinely representative *Comisiones Obreras*.

Another, more dangerous way of creating authentic space was by setting up parallel but unlicensed institutions. In the 1970s, for instance, the Polish opposition set up the so-called Flying University, a clandestine higher-education project in which banned texts were studied and the 'real' history of Poland taught. Almost all families had been touched by that history; the experience of being able to discuss with strangers and scholars the role of the Soviet Union in crushing Poland in 1939, in the Katyn massacre or during the Warsaw Rising was overwhelming but also irreversible.

In 1980, a small illegal group called the Free Trade Unions of the Coast opened up another space, the authentic trade union which took the name Solidarity. This space, opening at Gdansk, was soon replicated everywhere, first in big factories and then small ones and then in every office and workplace until the spaces began to run together into the shape of a free country. But this was not just a Braveheart freedom. Poland has always lived with the fear of being no more than the international space between Germany and Russia, a space which internal weakness or disorder could transform into one of those vacuums which powerful neighbours rush to fill. In 1980, everyone was aware of this risk. In 1981, General Jaruzelski's coup expressed his pessimistic view that authenticity in this particular space would always end in lethal disorder.

Solidarity produced authenticity in several flavours, some of which are not available even in Britain today. The union negotiated a list of guaranteed civil rights out of the regime, applicable to the whole country. But I am interested in Solidarity's own ideas about what made a space authentic, a site in which real people did real things.

The first condition was that workers should choose their own representatives, free of intimidation or constraint on expression. The second, much less well remembered, was that working people should take charge of their own working lives. Part of the fraudulence of daily experience had been the fiction of a workers' state, complete with dummy charters of shop-floor rights, although in reality the employees had no voice, no job security and no rights over their own labour. Solidarity therefore insisted on a new social-economic democracy of self-management, in which every enterprise would be owned and run by its own workers.

None of that remains in Poland, let alone in Britain. And yet that space to breathe is desperately needed. At home or in the shopping-mall, the consumer-citizen is loaded with all the regalia of fake sovereignty. At work, though, new management styles (flexibility, de-layering) enforce abject individual submission. Instead of a share in control, there is pretence consultation. Instead of unions, there are creepy staff associations. The word 'excellence' is a label for de-skilling programmes, for the move toward all-purpose cheap labour. Nothing is real; everything in this hyperwork culture is fraudulent and suffocating. It is time to start digging under Britain, to fashion spaces in which we can breathe, plot and be ourselves.

FURTHER READING

Anderson, Benedict, *Imagined Communities: Reflections on the Origin and Spread of Nationalism*, London: Verso, 1991.

Ascherson, Neal, *The Struggles for Poland*, London: Michael Joseph, 1987.

Ascherson, Neal, *Black Sea: The Birthplace of Civilization and Barbarism*, London: Vintage, 1996.

Ascherson, Neal, *Stone Voices: The Search for Scotland*, London: Granta, 2003.

Brody, Hugh, *The Other Side of Eden: Hunter-Gatherers, Farmers and the Shaping of the World*, London: Faber and Faber, 2001.

Garton Ash, Timothy, *History of the Present: Essays, Sketches and Despatches from Europe in the 1990s*, London: Penguin, 2000.

Hartog, François, *The Mirror of Herodotus: The Representation of the Other in the Writing of History*, trans. J. Lloyd, Berkeley: University of California Press, 1988.

Hobsbawm, E. J., *Nations and Nationalism since 1780: Programme, Myth, Reality*, Cambridge: Cambridge University Press, 1992.

Naipaul, V. S., *Beyond Belief: Islamic Excursions Among Converted Peoples*, London: Little, Brown, 1998.

Zimmerman, Andrew, *Anthropology and Antihumanism in Imperial Germany*, Chicago: University of Chicago Press, 2002.

7 Exploring space

JEFFREY HOFFMAN

Humans and machines: collaborators in exploration

'Exploration' at its most basic level is everything that expands the realm of human experience and of human consciousness. Although I spent nineteen years of my life as an astronaut, I readily acknowledge that the vast majority of space exploration is and always will be performed by machines. Therefore, it is important to understand how our unique biochemical-based minds can interact with machines to project our consciousness to realms we cannot physically reach. Exploration does not always require the physical presence of the explorer. I started my professional career as an astronomer, and I have always considered astronomy to be exploration, with telescopes the vessels that carry our minds, and thus at least part of our consciousness, out to the stars. Astronomy may be passive compared to actual physical exploration; but while Columbus with his ships gave us a new continent, Galileo with his telescopes gave us whole new worlds – and he never set foot outside Italy. During our lifetime, we have explored the surfaces of all the worlds Galileo discovered and many others besides without having physically travelled any farther than our own Moon.

The Hubble Space Telescope (see Fig. 7.1) is the flagship of our astronomical exploration fleet, with many powerful colleagues on the ground as well as partners in space observing other electromagnetic wavelengths. Hubble has taken us deep into space and back in time to the youth of our universe. The famous Deep Field photographs, revealing a multitude of previously unseen faraway galaxies, were taken by having Hubble observe tiny parts of the sky that were empty as far as other telescopes were concerned. If this is not exploring space, then what is? As our intellects travel far away with the ever-increasing gaze of our telescopes, our emotions and our spirits – also important aspects

Space: In Science, Art and Society, edited by F. Penz, G. Radick and R. Howell.
Published by Cambridge University Press. © Darwin College 2004.

FIGURE 7.1 Hubble Space Telescope (NASA photograph).

of our humanity – follow. Who is not moved by Hubble's images of the Eagle Nebula, where we can see creation itself? This is the biblical 'Let there be light!'

Of course, when we explore with telescopes, we are exploring passively and remotely. Nevertheless, for minds confined to a minuscule volume of space on a single, small planet to have expanded their domain to contemplate a universe as immense as the one we think we live in is an astounding feat of exploration. I do not propose to dwell on astronomy's role in space exploration except to mention my own good fortune in having played a role in repairing the initially flawed Hubble Telescope (see Fig. 7.2). Having been both an astronomer and an astronaut, for me one of the most satisfying aspects of working on Hubble was uniting the world of space astronomy, which normally prefers automated spacecraft, with the world of human spaceflight.

Despite my statement about the importance of machines in exploring space, I feel strongly about the value of human presence, and in what follows I shall discuss the human dimension of space exploration. But I do not want to do this in the context of man versus machine or humans versus robots, which I consider to be a false conflict. Robots, like all other machines, are our tools. We use them to expand our physical capabilities and to operate remotely in places we cannot go ourselves. The relationship of humans and machines has evolved throughout history as our machines have become more sophisticated,

FIGURE 7.2 JH and Hubble (NASA photograph).

and this evolution will continue with great benefits for the exploration of space, in which telescopes, satellites, probes, robots and humans all have roles to play.

Floating between heaven and Earth

The first question that most people ask me when they find out that I am an astronaut is, 'What's it like in space?' It is a very human question that we would never think of asking a machine, even though automated space probes do a superb job of measuring temperatures, pressures and many other aspects of their surroundings. But this is not what people are asking. They want to know, 'What is it like to *be* in space?' Probes send back data, but only humans can share experience. An important part of the tradition of exploration is to report on your travels and your discoveries, to share them with other people and ideally to make the new territory you explore part of general human culture and human consciousness. In this tradition, then, let me try to share some of my experiences in this new environment.

FIGURE 7.3 Shuttle launch (NASA photograph).

My first space flight, in April 1985, made me the 162nd human space traveller, of which there were 400 as of April 2001, the 40th anniversary of the first human spaceflight. This is a large enough number of people to make going into space perhaps no longer cutting-edge exploration in the sense it once was. Nevertheless, anyone sitting on top of a loaded rocket ready to make their first

FIGURE 7.4 Floating above Hubble (NASA photograph).

trip into space can rightly feel that they are embarking on a personal journey of discovery. After all, you can explore a new environment for yourself even if you are not the first person to go there. So, what do you see, feel and do when you go into space?

First of all, it's an incredible ride. I hope someday we will have gentler and safer ways of getting into space, and I will discuss this more below; but if you like excitement, the adrenalin rush doesn't get any better than feeling the boosters light and watching the ground fall straight away from you (see Fig. 7.3). You break the sound barrier in 45 seconds going straight up and in another minute the big blue sky of Earth has turned into the blackness of space.

The first thing you notice when you look back at the Earth is the new perspective. From a commercial jetliner you can look out the window and see entire cities below you. From an orbiting spacecraft you can see entire countries or even continents. I will never forget flying over Houston in the middle of the night during my mission to the Hubble Space Telescope, 600 km above the Earth, which is as high as the Shuttle ever goes. I could see the lights of Los Angeles on the Pacific coast out of one window and the lights of Miami on the Atlantic out of another. One of the most memorable views occurred when we installed new magnetometers and had to go all the way to the top of the telescope, 15 metres above the shuttle (see Fig. 7.4). Floating between heaven and Earth, this was one of the most spectacular and emotional moments in all my space flights. Attached to the arm by a slack, stainless steel cable, I could let go for a few moments and become a free-flying satellite. If my back was turned to the Shuttle, I felt truly alone in space.

Pictures can help give some idea of what space looks like, but the more subjective question of 'What does it *feel* like in space?' is much harder to communicate. I remember vividly my initial feeling of euphoria as soon as the engines shut down after my first launch. After all those years of waiting I was actually in orbit, weightless, in space at last. I looked out the window to see how fast we were going, but with no wind noise and no vibration there is no real feeling of speed. Still, when you fly over Boston and Cape Cod and then watch Land's End go by ten minutes later on the other side of the Atlantic Ocean, you know you are going fast!

There are some very strong physical sensations during the initial experience of weightlessness. It is a bit like being upside down on Earth. Without gravity pulling your blood toward your feet, the blood rushes to your head. Hanging upside down for several hours or several days can give you quite a headache. Releasing gravity's constant pressure on your spinal column increases your height by several centimetres, which may be ego-boosting for someone who on earth is just a little bit short of being a six-footer, but the physical effect on your back can be similar to what it must have been like to be on a medieval torture rack. Then there is space sickness, resulting from a total confusion of your body's vestibular system, which is responsible for maintaining your balance on the Earth and which suffers from the lack of the expected gravity-induced cues for up and down. We can treat most of the unpleasant symptoms people suffer when they enter space, but the initial accommodation to weightlessness involves significant stress.

Humans are remarkably adaptable, though, and after a few days in space everyone gets over the initial problems of adapting to this new environment and starts enjoying the experience of actually being in space, floating weightless, looking at the world beneath and space all around. Everyone reacts in his or her own way, but I have found that being in orbit almost universally evokes a sense of profound awe, in the deepest sense of the word. It is not dissimilar to what many people feel in the high mountains, where you have to work hard in an unforgiving environment and take certain risks to gain the summit. On the high ground, your view is expanded. Your mind is cleared. Space is much higher than the highest mountain-top, and the environment is even more unforgiving. Realising that outside your spacecraft or your spacesuit you would not survive more than fifteen seconds, you feel all around you the extreme hostility of most of the universe toward life. Then you look down at the Earth. Seeing the thinness of our atmosphere and realising that this is all that separates us from the hostility of space may make you may feel a new regard for the fragility of life on Earth.

The most powerful and unique feeling in spaceflight, however, is the physical freedom of weightlessness itself, which is an utter delight. For me, the experience is more than just physical; it has psychological, emotional and even spiritual dimensions. We don't possess the shared experience to express this adequately through language, and even pictures can't convey the inner feeling of weightlessness. One of my crewmates, an amateur juggler, told me that throughout history the dream of jugglers has been to create the illusion that they could make balls appear to hover motionless in front of them. In space, you can do this – it's no illusion. Have you ever dreamed of flying? Again, in orbit you can do it (see Fig. 7.5). Weightlessness gives you a wonderful sense of freedom, an ability to do outrageous things barely imaginable back on Earth. The joy of weightlessness is one of the main reasons I believe that humanity has a future in space.

The work of astronauts

At the end of this chapter I will discuss whether and how large numbers of people might someday get to explore space personally, but at this point I need to emphasise that NASA does not send me or my colleagues into space so that we can see beautiful views or experience the physical bliss of weightlessness. They send us to do serious work, which in my case has taken advantage of the many capabilities of the Space Shuttle. We have used Shuttles to carry satellites into orbit, to repair the ones that are there and to retrieve broken or misplaced satellites and bring them back to Earth. We have used Shuttles, especially when outfitted with the European-built Spacelab, to conduct many experiments in the life sciences, fundamental physics, crystallography and many other areas of research in fields where gravity is an important experimental variable. We used the Shuttle as a crew transport and logistics supply vehicle for the Russian space station Mir, adding many years to Mir's useful life. Now we are

FIGURE 7.5 Magic carpet (NASA photograph).

FIGURE 7.6 International Space Station (NASA photograph).

using the Shuttle to build and maintain the new International Space Station (see Fig. 7.6).

I will not discuss the new space station in any detail, since our subject is exploring space; and what astronauts are doing in low-earth orbit these days is not so much exploring as learning how to live there comfortably and take advantage of this unique environment, while at the same time trying to figure out how eventually to explore further away from the Earth. Scientific experimentation is one of the primary functions of the space station, although by no means the only one. The station's most important use may actually turn out to be as a testbed for new space technologies. For instance, more efficient solar power generation systems or deployment mechanisms for large foldable antennas could be tried out for possible installation on new, expensive satellites.

But of all the things the new space station can become, probably the most important is a testbed for new space robotic technology. Robotic technology will be absolutely essential to allow human beings to utilise fully the opportunities presented by the station. There are so many experiments and maintenance requirements and so few people. And robotic technology will be crucial in allowing us to explore further away from the Earth. The space station is the

FIGURE 7.7 Robonaut (NASA photograph).

place to test and perfect robotic servicing techniques that will eventually enable repair and maintenance of what we expect, in a decade or two, will be fleets of astronomical, solar and earth-observing satellites, parked at stable positions over a million kilometres from the Earth – much as we astronauts serviced Hubble.

After we repaired the Hubble Space Telescope, I was sometimes asked whether robots could not have performed the repairs as well as we did (see Fig. 7.7). The answer was and still is 'no'. Neither then nor now are space robots sophisticated enough. They simply do not have sufficient tactile capability to do a lot of the fine work that we carried out, such as dealing with tiny non-captive screws on the back of electronic connectors. These screws came loose and ended up floating all over like a swarm of flies. Also, we had several surprises that forced us to adopt procedures completely different from those planned before the mission or suggested in real-time by Mission Control – for example, in figuring out how to close a warped door, which would have crippled the telescope if left open. In unplanned, 'time-critical' situations, humans are far more efficient and flexible than even our best robots. However, I hope that this will not always be true. As I have emphasised, robots are tools, and the better they become the

FIGURE 7.8 Mars rover (NASA photograph).

more useful work we can do with them. If we do not eventually have robots capable of the sort of repair work that we carried out on Hubble, that will mean that progress in space robotic technology has stalled out – a huge pity for space exploration, since, no matter how much progress we make in human spaceflight, there will always be places we cannot go ourselves physically, and the better robots we have, the more sophisticated exploration we will be able to carry out.

In praise of machines

I love robots in space! They have taken me as close as I will ever get to the surface of Mars. When I watched the 1997 Martian Pathfinder rover moving around, poking up against rocks and finding its way around obstacles, I felt the presence of human intelligence on the surface of Mars (see Fig. 7.8). While tools were originally limited to enhancing human physical capabilities, modern technology has made it possible to use tools to project human presence. Talking about robots competing with people in exploring space does not make any more sense than talking about robots competing with telescopes. Humanity's most distant physical explorer is Voyager 1, now over 10 billion kilometres from the sun, about 11 light hours away or about 0.03% of the distance to

Alpha Centauri. Comparing this to the most distant objects observed by our telescopes (over 10 billion light years away) we can say that our robotic probes have access to less than one part in 10^{29} of the volume of the universe accessible to our telescopes. We can dream of sending probes to the stars. Indeed we are working on technologies that may someday enable us to do this; but barring the realisation of science fiction space warp technology, almost the entire volume of the universe will have to be explored with telescopes, not with robotic probes.

Inside our solar system, telescopes and other remote-sensing instruments in orbit around celestial bodies can give us a much closer look than our Earth-based telescopes; and machines that can manipulate objects on the surface of another planet give us tremendous power to explore – far beyond just looking or even taking in situ measurements. Good vision is essential, but touch and manoeuvrability take us into another dimension of control over the environment being explored. In places robots can get to, they can do things that telescopes cannot. There is nothing terribly complicated about this. In places people can get to, they can do things that robots cannot do. It is just harder, riskier and more expensive to send people than robots, just as it is harder, riskier and more expensive to send robots than to look with a telescope. The farthest away from home that human beings have ever explored is the Moon, less than half a million kilometres away. Even when we someday get to Mars, human beings will still have reached only a tiny fraction of the domain already accessible to our robotic probes, just as our robots have reached only a fraction of the domain explored by our telescopes. That is the reason for my statement at the beginning of this chapter that most space exploration has been and always will be done by machines, meaning telescopes, satellites, probes and robots. But where people have gone, we have done things that machines could not do, and we have experienced these environments as human beings.

Communicating the experience

Granted that humans can make a contribution in carrying out scientific experiments in space, we also need to remember that there are other types of knowledge besides scientific; and human presence is vital to communicating the experience of new environments in ways that go beyond science. Few people have actually travelled in the Arctic or the Yukon, but most of us have some idea of what these Arctic regions are like. How much of our general perception of the Arctic comes from scientific journals, I wonder, and how much from

FIGURE 7.9 Hadley Rille (NASA photograph).

literary accounts, such as Jack London's famous short story, 'To light a fire'? Exploration goes beyond science. Truly to 'know' an environment in human terms, and especially to communicate what that environment is like, requires art as well as science, and it helps to have been there. Alan Bean landed on the Moon during the Apollo 12 mission in late 1969. Ten years later, Alan was the mentor of our group of young astronaut recruits, and I was impressed that this Navy fighter pilot enjoyed talking about literature and art. He eventually left NASA to try his hand as a full-time artist, saying that he wanted some way other than words or photographs to try to share his experiences. Most of the lunar landscape is well represented by black-and-white photographs, with the only colour coming from human artifacts. In painting Hadley Rille – a long, sinuous, steep-walled canyon, often called the most spectacularly beautiful lunar feature to be visited during the entire Apollo programme – Alan tried to express the rich variety of surrounding surface textures through an almost impressionistic use of colour (see Alan Bean's painting on the back cover). The Moon doesn't necessarily look like this, any more than the Rouen cathedral that Monet painted so many times actually appeared in all those different colours. But for me, Alan has shown something of what it is like to be on the Moon that I would never have been able to experience just from a photograph.

Now, robots may not write books or paint pictures, but as I said there will always be places our robots can go where our bodies cannot follow. So if exploration is ultimately the expansion of human consciousness, then we need to do everything we can to enable our minds and spirits to follow our robots where our physical bodies can't go. In future space exploration we will need to use all the virtual reality tools that we possess to put ourselves inside our robots – to use robotic presence to extend our sensory experience – because our consciousness of the world is intimately tied to our sensory experience of the world. With vision, we can already do pretty well. Given proper sensors on the lunar surface and a fast enough data link, I could stand on the Earth, look around, and have almost the same visual experience that Alan Bean had on the Moon. Sound should not be a problem if we are in an environment where this is important. Future Mars probes will carry microphones onboard which will let us hear the natural sounds of wind and dust on the Martian surface and the sounds of the probes digging up Martian soil.

What about taste and smell? There is a passage in Norman Mailer's book about the Apollo programme *Of a Fire on the Moon*, in which he says that ultimately to know the Moon as human beings we need to be able to smell a lunar rock. This is a nice thought, and I would love to try it. In general, however, taste and smell play such a small role in human society that we have never bothered to develop technologies to transmit these senses over a distance, and I do not see the lack of virtual smell and taste capability as a significant barrier to space exploration. The fifth sense, touch, is something else, however, because the ability to manipulate the environment we are exploring is critical. To really expand human presence into a new environment, we need to incorporate sophisticated haptic – that is, touch-based – feedback systems into our space robots. With sufficiently sophisticated visual and haptic feedback, I can imagine being able to put robots on the Moon and having back on Earth the experience not just of seeing but actually of exploring the lunar surface. The $2\frac{1}{2}$-second round-trip light travel time is short enough that, with sufficient training, you could guide a robotic partner around the surface of the Moon almost as though you yourself were there on the spot. You could look under rocks, dig for soil samples, and you would eventually come to feel that a significant fraction of your consciousness was present on the surface of the Moon. That is the nature of virtual reality, as described by Char Davies in her chapter on 'Virtual space': to create a convincing internally perceived environment. Such virtual-reality-equipped robots would be incredibly powerful tools for lunar exploration.

Extrapolating to Mars, however, not to mention farther out in the solar system, forces us to face the problem of the finite speed of light. We can transmit back sufficient information from Mars to permit reconstructing a virtual presence after the fact, so that using a computer cursor you can fly virtually through Martian canyons. However, the idea of interactive virtual presence changes fundamentally as you increase the communication travel time. Flying a computer cursor through a virtual landscape is one thing, but with a roundtrip travel time of tens of minutes could you control a real airplane flying through Martian canyons from a computer located on the Earth? From a strictly technical point of view, the best current solution is to build more intelligence and autonomy into our robots so that they can do a maximum of useful work on their own while waiting to communicate with us. However, the degree to which autonomous robots far away in the solar system can give us a real conscious awareness of new environments will be far less than totally interactive real-time virtual reality, and of course the speed of exploration will be enormously slower.

Destination: Mars

I do not foresee the speed-of-light travel delay as a long-term impediment to exploring Mars, because eventually we will have humans on Mars. I cannot predict when, but Mars seems such a fascinating place that once we have the capability to send humans there for a reasonable cost and with acceptable risk I have no doubt that we will do it – in large part because the results of our telescopic and robotic exploration will be so exciting that people will feel that they want to get to know this planet as well as is humanly possible. That is the key, 'as well as is *humanly* possible', because ultimately the best vessel to carry human consciousness to a new environment is the human body. When we go, however, we are not going to throw away our robots. Just the opposite: human explorers on Mars will exercise real-time control over armies of exploration robots, mostly autonomous, but susceptible when necessary to real-time assistance to overcome problems beyond the capability of their artificial intelligence to solve. We do not necessarily have to develop robots as good at complex decision-making as humans for them to be useful explorers, especially if a human explorer is nearby to redirect them when they do not know how to proceed, and to fix them when they break. Imagine what one human explorer could do with a flock of robots even as smart as, say, the average dog. What I am looking forward to is in a sense a symbiosis between human and robotic exploration. Each makes the other more efficient, and the closer together they are physically the more opportunities there are for symbiosis.

One of the strongest lures for exploring Mars is the ever-growing indication that Mars in the past may have been very different from the dry, airless desert it is today. Learning the detailed history of our neighbouring planet will someday occupy the full attention of as many field geologists as we can support on the surface of that planet. It seems as though every month we are surprised yet again by new evidence that liquid water may once have flowed on Mars. Analogies are not evidence, but everywhere on Earth that we have liquid water, we have found life. If life once existed on Mars, what happened to it as temperatures dropped, the atmosphere dissipated and the surface dried up? Did it leave fossils? Did any of it survive anywhere? It would be wonderful if the currently planned generation of Mars robotic explorers could find evidence of organic material near the surface. What an impetus for further exploration if they succeed! But if they do not, I hope we will not give up the search.

It is interesting to speculate about what life on Earth would do if the same changes we think happened on Mars happened here. Life exists in some pretty hostile environments on the Earth, including several kilometres underground, where some enterprising bacteria have actually learned to metabolise rock. What would happen if the Earth's oceans evaporated and our atmosphere disappeared, with the loss of the greenhouse effect turning our planet into an arctic desert that would make current-day Antarctica seem like a tropical paradise? If this happened, the most likely life forms to survive may in fact be just those hardy subterranean bacteria. Our Martian robots will certainly dig exploratory holes, perhaps as deep as several metres, maybe even deeper; but I suspect if we want to dig really deep our field geologists on the Martian surface will have to be joined by some drilling-rig roughnecks. Once the scale of human activity on Mars gets large enough, we will need support personnel as well, to maintain the scientific bases and take care of all the logistical activities that will be necessary to allow human survival in such a hostile environment. That is the situation now in Antarctica. But in addition to extending vastly the range of scientific exploration that can be carried out on the surface of Mars beyond what robots could do on their own, I hope that at least one of these human explorers turns out to be another Jack London or Alan Bean, able to give the rest of us, who will never go there, a sense of what Mars is really like. Not just the data, not just the views, but a sense of what it is like to *be* on Mars.

Some challenges ahead

In imagining future human exploration outposts on Mars and the Moon, similar to the scientific research stations we currently maintain in the Antarctic, I am assuming that human beings can tolerate long-term exposure to partial gravity better than they tolerate large periods of weightlessness, or 'zero g'. But we do not know that for a fact. Up to now, almost our entire long-term experience with gravity has been either at one or zero g. Gravity is not a binary phenomenon, however. A tremendous amount of research has been devoted to the physical effects of zero g, such as bone loss, cardiovascular deconditioning, muscle atrophy, fluid shifts, depression of the immune system and changes in blood cells. Lunar gravity is about one-sixth the Earth's and Martian gravity is a little over one-third. Is there a gravity threshold above which certain problems disappear? Are these thresholds different for different phenomena? Does the severity of the problems scale linearly with gravity, or exponentially? It is humbling to admit that after forty years of human space exploration, we do not have a clue. Astrobiology, which NASA talks a lot about these days, is not just the search for extraterrestrial life. It is also the study of the response of terrestrial life to non-terrestrial environments. Not just human life, but all aspects of life down to the cellular level, where we have tantalising indications that even something as basic as the expression of individual genes can be affected by gravity. The first opportunity we will have to study in detail the response of living systems to extended partial gravity levels will be with a centrifuge in orbit, such as the one that may someday be installed on the new International Space Station. This is one reason why I referred earlier to the station as a place where we can examine strategies for exploring further away from the Earth.

When we have the technology, the economic means, and the political will for humans to explore Mars, I am sure that at the same time we will be working on advanced robotic exploration of the outer solar system and perhaps beyond. After all, if our telescopes find Earth-like planets around a neighbouring star, won't we want a closer look? Developing the artificial intelligence and self-repair capabilities required for a robotic interstellar probe will be daunting challenges, although in a sense the technical challenges may turn out to be easier than developing a civilisation with sufficient social stability and foresight to undertake a scientific research project that may take decades, or even centuries, to complete.

In any case, the parallel challenges of future space exploration will be to push the outer boundaries of robotic exploration into realms previously explored only by telescopes and at the same time to push the boundaries of human presence into realms previously explored only by robots. A metaphor for exploring space might thus be an expanding sphere of human consciousness and experience composed of three layers. On the outside is the overwhelmingly large part of the universe accessible only to passive sensing. Tens of orders of magnitude smaller is the part of the universe that we can physically interact with through autonomous and remotely controlled machines, with the level of virtual human presence limited by the speed of light. And in a volume smaller by tens of orders of magnitude still will be that part of the universe that humans can experience directly, with no speed-of-light compromises. How fast these three layers will grow will depend on many factors, most of which we probably can't even imagine today, much less be able to predict. One limiting factor which we can clearly identify right now, however, is the inordinately high cost and risk of getting anything into space, machines or humans.

This brings me back to whether large numbers of people will ever be able to explore for themselves the new environment of space. I have mentioned virtual presence as an important aspect of future human space exploration, but I doubt that virtual reality will ever be able to recreate for people on the Earth the total experience of going into space – unless someone can figure out how to build a gravity shield to allow you to experience weightlessness in your living room. A gravity shield would utterly change space travel, not to mention many other aspects of our industrial civilisation. Unfortunately, our current understanding of physics seems to indicate that gravity cannot be shielded, and unless this turns out to be wrong, then to find out what it is really like to be in space, you will have to go there.

Space tourism: when?

Getting into space is currently too expensive and too dangerous to be accessible to enough people to really be called 'tourism'. It is not outrageous, however, to imagine that this will not always be true. Go back a century to the heroic age of Antarctic exploration, when only legendary adventurers such as Amundsen, Scott, Mawson and Shackleton, travelling on ships, dogsleds and skis, could penetrate the Antarctic continent. It was dangerous and expensive. Many people died. It would have seemed improbable to these explorers at the dawn of the twentieth century that in a little more than fifty years we would have a permanently inhabited scientific outpost at the South Pole; and it would have seemed outrageous that by the end of the century significant numbers of tourists would visit there. But air travel made it possible, and thousands of paying, private tourists now visit Antarctica every year, some of them going as far as the South Pole.

How long will it take before an equivalent revolution occurs in space travel? We already have the technology to build a hotel up in space; it would merely be a more luxurious version of a space station. The problem is that people cannot get there, at least not now. I mentioned the excitement of riding a Shuttle into space, due in large part to the enormous amount of energy that has to be harnessed just to get off the ground. Is this much energy really necessary to go into orbit? A simple physics calculation shows that the potential and kinetic energy gained by a mass of 1 kilogram when it is lifted 300 kilometres above the surface of the Earth and accelerated to an orbital velocity of 8 kilometres per second is about 35 million Joules. This sounds like a large number, but a chemical thermodynamics table shows it to be equivalent to the energy of combustion of about 700 grams of propane, or in terms of electrical units about 10 kilowatt-hours. We may complain about our gas and electric bills, but the utility companies charge a lot less for this amount of energy than the approximately $20 000 that is usually quoted as the current cost of putting a kilogram into orbit. We should be able to do better.

Part of the problem lies in the rocket equation. At present, we fire our engines the whole way into orbit, and the fuel that pushes us during the latter part of the ascent has to be carried all the way up during the earlier part, which eats up fuel exponentially. Imagine the savings if we could power an ascent externally. Could we shoot some things needed in space from the surface of the Earth using a high-velocity electromagnetic rail gun? Of course, the enormous initial acceleration would not be practical for humans, but think of the enormous quantities of water, fuel, food, building material and other supplies needed to sustain life on a space station or support a trip to Mars, all of which can quite nicely tolerate very high g-levels. Once underway, could the payload be given additional external energy from high-power lasers? A lot of expense could be saved if all this material could be put into orbit without suffering the inefficiency of the rocket equation.

More gentle transportation systems would be necessary, of course, for humans and other delicate cargo. What forms these 'gentler' systems may take is the subject of considerable technological research and a lot of science fiction. I don't know when it will happen. But once access to space is affordable and safe enough to allow large numbers of people to explore for themselves the new world of Earth orbit, and the first real space tourists come back with their travellers' tales, you won't be able to keep people away.

Space tourism would be a huge boon for space exploration, because the existence of technology funded by the private sector and capable of getting sufficient mass into low-Earth orbit to support significant human presence will make possible human and robotic activity on the Moon, Mars and beyond at costs far less than we face today. This is fortunate, since almost surely the development of space will in the end be driven by economics more than by exploration. I am not going to speculate further about these developments; but wherever space tourism may ultimately lead, one thing about human space travel is sure. Until elderly people decide to throw away their canes, walkers and wheelchairs and move to extraterrestrial retirement homes to finish their lives with a physical freedom that they thought they had lost forever, everyone who ventures into space will, with luck, come back. So let me finish by sharing some of what that experience is like.

Homeward bound

When the moment comes for you to fire your retrorockets and start your plunge into the atmosphere, it does not matter how long you have been in space. You face the loss of the beautiful sights you have seen and of the precious freedom of weightlessness. No matter how much I looked forward to seeing my family and friends after a flight, at this moment I always felt a slight sense of sadness.

When the engines fire, they only produce a tenth of a g, but after being used to weightlessness, you feel like the acceleration is slamming you against the wall. The burn only lasts a few minutes, and then you start your free fall toward the Earth, enjoying a few last minutes of floating (how long depends on how high your orbit is) until you hit the atmosphere at Mach 25, and soon the friction that will eventually slow you down to landing speed starts to heat the atmosphere around the Shuttle.

What starts out as a dull glow, visible only if your reentry is at night, soon turns into a magnificent light show, with the confluence of superheated, glowing shock waves trailing behind the Shuttle like the wake following a motor boat. For a short time you are a meteor, or perhaps better to say 'meteorite', since the idea is to survive the entry and land safely on the Earth. The light show fades away after around ten to fifteen minutes, and soon you are in the sky over Florida. There is a vibration as you break the sound barrier for the second time in your flight, slowing to your subsonic landing speed. The pilot is doing most of the work now, circling to line up with the runway, and pretty soon you feel the wheels hit the ground and it is over. Looking at what seems to be an airplane touching down, it is hard to believe that this Shuttle and its crew were in space half an hour ago. But now you are on the ground, and you feel very heavy.

I mentioned being asked the question, 'What is it like in space?' I will finish with another of the most frequent questions that people ask me about space travel: 'Have you somehow been changed by the experience of being in space?' Again, this is a very human question, meant to be answered by another human being. In the short term, space flight produces a lot of changes. I remember moving my head around while unstrapping from my seat and feeling like my head was a tennis ball bouncing from one side of the cabin to the other. When I got up and started to walk around, aside from feeling very heavy, I found I was lurching a bit like a drunken sailor every time I went around a corner. Your vestibular system takes a while to readapt to gravity, and the longer you have been in space the longer it takes. Your other physical systems take some time to readapt as well, but eventually everything seems to get back to normal.

Psychological changes, however, are what people seem to be most interested in when they ask this question. Some mental changes can lead to humorous moments for first-time returning space travellers who don't know what to expect, as I learned during my first meal back on Earth, when I left my fork floating in front of me – or so I thought – while I reached for my drink. But what about more fundamental changes? Are you changed by experiencing great beauty? Are you changed by a totally unique physical experience? How does your life change after any peak experience? In trying to give a human answer I will turn again to art, to a poem by the French surrealist René Daumal, from his book *Mount Analogue*. The poem has circulated for years within the American mountaineering community but goes beyond mountaineering and speaks to the heart of all exploration. I have carried a copy on all my space flights to share with my crewmates shortly before returning to Earth:

> You cannot stay on the summit forever;
> you have to come down again . . .
> So why bother in the first place?
> Just this:
> What is above knows what is below,
> but what is below does not know what is above.
> You climb, you see.
> You descend, and you see no more,
> but you have seen.
> There is an art of conducting yourself in the lower regions
> by the memory of what you saw higher up.
> When you can no longer see, you can at least still know.

FURTHER READING

Chaikin, Andrew, *Space: A History of Space Exploration in Photographs*, London: Carlton, 2002.

Freeman, Marsha, *Challenges of Human Space Exploration*, Chichester: Springer-Praxis, 2000.

Kelley, Kevin W. (ed.), *The Home Planet*, New York: Addison Wesley, 1988.

Reichardt, Tony (ed.), *Space Shuttle: The First 20 Years – The Astronauts' Experiences in Their Own Words*, Washington, DC: Smithsonian Institution, 2002.

Sagan, Carl, *Pale Blue Dot – A Vision of the Human Future in Space*, New York: Random House, 1994.

Stevens, Payson R. and Kelley, Kevin W., *Embracing Earth: New Views of Our Changing Planet*, San Francisco: Chronicle Books, 1992.

Stoker, Carol P. and Emmart, Carter (eds.), *Strategies for Mars: A Guide to Human Exploration*, Science and Technology Series, **86**, San Diego, CA: American Astronautical Society, 1996.

8 Outer space

JOHN D. BARROW

Space is blue and birds fly through it.

Werner Heisenberg

Introduction

In 2000, we were bombarded with information about 'time'. If you attended the Millennial Darwin Lectures you will undoubtedly have been told by some cosmologist that 'time' exists to stop everything happening at once. Now it is the turn of 'space', and you will learn, likewise, that 'space' exists to stop everything happening in Cambridge. Fortunately, space is superficially a little less mysterious than time – it does not have a special direction and we don't have a pervasive subjective sense of its passing. There do not seem to be a number of different spaces in the way that there might arguably be different measures of time. None the less, for the physicist space and time are rather more alike than others might think. Where once we talked about space and time, now we focus upon the united front of 'space-time'. The existence of an unsurpassably special quantity in the universe with the units of a velocity – the velocity of light in a vacuum, 299 792 458 metres per second – ensures that there is an indissoluble link between space and time.

Ancient ideas about space appear to be very complicated. They were pre-occupied with location: everything had its place and there was a place for everything. The only greater issues were those concerned with the existence of infinity or the void.

An exploration of modern ideas about space involves an exploration of mathematics, physics and astronomy. All three have become intimately entwined as

Space: In Science, Art and Society, edited by F. Penz, G. Radick and R. Howell.
Published by Cambridge University Press. © Darwin College 2004.

a result of the discovery that space is not absolute and flat, a mere arena for objects to act out their parts. Space is seen to be dynamic, flexible, curved, with a history and a geography whose potential complexities we are just beginning to appreciate.

Space is not a static stage

For millennia space was regarded by scientists and philosophers as a fixed background stage. The heavenly bodies moved upon this space without affecting its properties. It was the arena in which interactions between masses took place, governed by Newton's laws of gravity and motion. The laws of motion that scientists discovered were prescriptions for how objects would move around on the flat tabletop of space. Whatever happened to those motions nothing could change the shape of space or destroy it. Even if all the matter, motion and energy in the Universe were to disappear, space and time would still remain.

The great revolution in outlook that Einstein brought about was made possible by the work of nineteenth-century pure mathematicians like Bernhard Riemann, Nikolai Lobachevskii, Janos Bolyai and Carl Friedrich Gauss who systematised the study of geometries that were not Euclidean. Euclidean geometry is the map of flat surfaces where the interior angles of all triangles sum to 180 degrees. In many ways it is surprising that these developments took so long to come to fruition because artists and designers had appreciated the logic behind geometrical constructions on curved surfaces for hundreds of years.

Amongst mathematicians and natural philosophers there had long been interest in the problem of whether or not Euclid's geometry really needed the famous parallel postulate, that parallel lines never meet. It was widely believed that Euclid's geometry was the only geometry that was logically possible and this geometry described the way that space really was. It was not a model. It was not an approximation. It was part of the absolute truth of things. As such it was held up as a shining example of the ability of human reasoning to fathom one part of ultimate reality. And if it could do it in one place, why not in others as well? Thus, the theologian who was challenged that the quest for an understanding of ultimate truths was hopeless, could point to the truths of Euclidean geometry as a counterexample.

The discovery that there were logically self-consistent geometries other than Euclid's, in which the parallel postulate did not hold, was therefore rather shocking to non-mathematicians. Space was now relative, not absolute. There

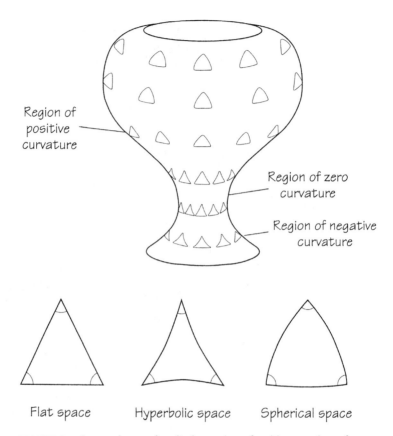

Flat space   Hyperbolic space   Spherical space

FIGURE 8.1 A vase whose surface displays regions of positive, negative and zero curvature. These three geometries are defined by the sum of the interior angles of a triangle formed by the shortest distances between three points. The sum is 180 degrees for a flat 'Euclidean' space, less than 180 degrees for a negatively curved 'hyperbolic' space, and more than 180 degrees for a positively curved 'spherical' space.

were an unlimited number of different logically consistent geometries which described the properties of triangles on variously curved surfaces. In Fig. 8.1 we can see an everyday object that has a surface displaying regions where the curvature of space is positive, zero (flat), and negative. The curvature of these surfaces is defined by the sum of the interior angles of a triangle formed by joining three points that are not collinear by the shortest allowed paths. If the sum of those angles is 180 degrees then the surface is flat (Euclidean); if it is

FIGURE 8.2(a)−(b) (a) *The Arnolfini Wedding* by Jan van Eyck, 1434. The entire scene is reflected in the small convex mirror hanging on the rear wall, which is shown in (b). © National Gallery, London.

FIGURE 8.2(b) *(cont.)*

less than 180 degrees, the surface is said to be negatively curved; and if it is greater than 180 degrees, it is said to be positively curved.

It is possible that this discovery could have been made a lot earlier than the early nineteenth century by the use of mirrors. If we look at the famous fifteenth-century painting of *The Arnolfini Wedding* in the National Gallery we see that there is a faithful representation of the foreground scene reflected in the curved mirror on the back wall (see Fig. 8.2b). What is seen in the curved mirror is a non-Euclidean geometry.

Since it was known that there was a one-to-one correspondence between an object and its image in the curved mirror, governed by the laws of reflection, perhaps it could have been deduced that there must have existed an axiomatic basis for the non-Euclidean geometry seen in the curved mirror? Look at Euclid's pictures in a curved mirror and you have the theorems of non-Euclidean geometry.

Space is wrinkly and curved

William Clifford was the first, in 1870, to suggest that the geometry of space might be not only curved, but changeable. He imagined ripples of curvature running through space that was only flat on the average, governed by equations akin to the continuity equations that govern the flow of a liquid.

176

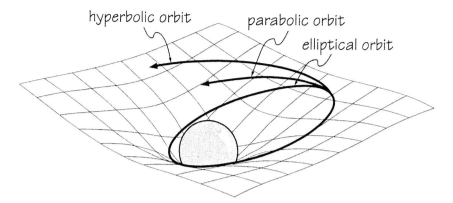

FIGURE 8.3 Einstein's conception of curved space. Bodies take the quickest route between two points on a curved surface.

Clifford had the right idea, but it took an Einstein to choose the right mathematical tools needed to forge the link between geometry and physics. The result, the general theory of relativity, is one of the most striking creations of the human mind and looks the same today as it did when it was published in 1915. It expressed precisely and testably the idea that the presence of mass and energy in space and time determines the geometry of space and the flow of time locally. There is no fixed background geometry of space. Geometry is determined by the local presence of mass-energy and its motion. As mass and energy moves so the geometry changes. There is a symbiotic relationship between mass, motion and the geometry of space encapsulated in John Archibald Wheeler's one-line summary of Einstein's theory: 'Matter tells space how to curve; space tells matter how to move.' Einstein's mathematical achievement was to supply the system of equations that allow us to determine the geometry that is created by any given configuration of matter (the generalisation of Newton's law of gravitation) together with the equations of motion that tell us how bodies move in this curved geometry (the generalisation of Newton's laws of motion). In Fig. 8.3 we see how this transforms our view of a planet orbiting around a star. Think of space as a malleable rubber sheet that is distorted by the presence of masses in it. The presence of the star curves the space in its vicinity and the planet takes the quickest path between any two points when acted upon by no forces. The Newtonian idea of bodies moving on a flat space subject to gravitational forces is superseded. Moving objects take

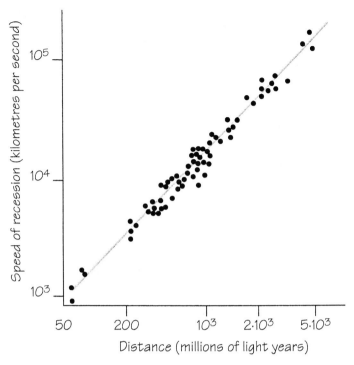

FIGURE 8.4 Hubble's law. The linear increase of the speed of recession of distant sources of light versus their distance from us found by Hubble and his successors.

their marching orders from the local shape of space, not from any mysterious force called 'gravity' which transmits its effects infinitely quickly. The farther an object is from the star the flatter the space will appear. The force of gravity has been replaced by the curvature of space.

At first one might think this is just another way of saying that there are gravitational forces but there are many differences from the old picture of absolute flat space. If we spin an object on a fixed table-top there will be no effects of the rotating object on other objects far away. But spin an object on a rubber sheet and the sheet is twisted around in the same direction, communicating the effects of rotation to distant objects, dragging them around in the same direction. The effect now observed is called the 'dragging of inertial frames'. Next, suppose that a very irregular sequence of pulsations is occurring somewhere on the rubber sheet. It will transmit undulations outwards through the sheet. These ripples in the curvature of space are called 'gravitational waves' and are expected to be emitted by any changing process in the Universe that is not perfectly spherical in shape, for example, two colliding stars. Only in the most violent circumstances will we have a chance of detecting them because gravity is so weak but new large detectors have begun to search for their effects when they impinge on the Earth. A third consequence of the rubber-sheet picture of space is that by accumulating a large enough mass in a small enough volume of space it is possible to pinch it off and isolate it from the rest of the Universe. Such a region is called a 'black hole'.

Space is big

The most dramatic consequence of Einstein's theory of curved space and time was its prediction that there is no static stage of space that provides a backdrop against which all the local meanderings of the stars and planets take place. The entire Universe is in a state of overall change and the sense of that change is expansion. By this we mean that if we could measure the separation of distant clusters of galaxies we would find their separation tomorrow to be greater than it is today. Nevertheless, we are not expanding, the Earth is not expanding, nor is our galaxy. These structures are able to withstand the expansion because they have a larger than average local force of attraction to counter it. Only when we reach the scale of great clusters of galaxies do we find the markers that participate in the expansion of the Universe. The first solutions of Einstein's equations which displayed this expansion were found in the early 1920s by the young St Petersburg mathematician and pioneer meteorologist, Alexander Friedmann. These 'Friedmann universes' still provide us with the best description of the overall structure of the Universe and its expansion of space in time.

Seven years after Friedmann's discovery, Edwin Hubble found convincing observational evidence that the Universe was indeed expanding. The spectra of light from stars in distant galaxies were shifted systematically toward the red end of the optical spectrum compared with the same spectral lines on Earth. There was a simple interpretation. It was an example of the well-known Doppler shift whereby waves emitted by a receding source are received less frequently than waves emitted by a stationary source. By measuring the amount of the frequency shift it was possible to deduce the recession velocity of the star from which the light was received. The identity of the line being shifted is known from its shape and location relative to other lines. By comparing objects of

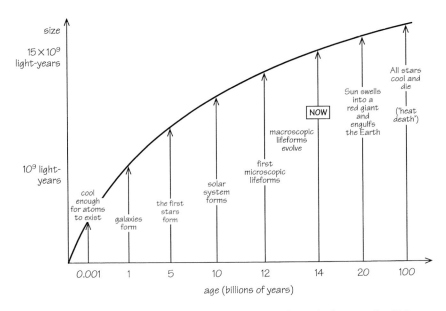

FIGURE 8.5 The increase of the scale of space with time in the expanding Universe, showing the chronology of development that occurs as the Universe expands and cools allowing matter to form atoms, molecules, galaxies, stars, planets and people.

the same intrinsic brightness it was possible to infer their distance from their apparent brightness. The result was Hubble's famous expansion 'law' showing the speed of expansion of the Universe increasing linearly with distance, see Fig. 8.4.

The expansion means that the average conditions in the Universe are continuously changing. The temperature and density of matter were greater in the past than today. Astronomers have gradually pieced together a pleasingly consistent picture for the universal history that unfolds as the Universe expands. It is shown in Fig. 8.5.

Modern astronomy has confirmed the expansion of the visible universe in some detail and revealed that we live about 14 billion years after the apparent beginning of the expansion. In order to reconstruct its history at very early times we need a full understanding of high-energy physics. When the Universe is younger than about 10^{-10} seconds, it experiences extremes of temperature and energy that exceed those we can produce on Earth in particle colliders at CERN (European Organisation for Nuclear Research) and elsewhere. However, when the Universe is between 1 second and 3 minutes old it behaves like a vast nuclear reactor producing definite abundances of the light isotopes deuterium, helium-3, helium-4 and lithium-7 which can be calculated in detail and correspond to the abundances observed in the Universe today. Those detailed astronomical observations allow us to confirm the consistency of our model of the Universe right back to these very early times. Likewise, the very existence of the microwave background radiation, its spectrum – indicative of pure heat radiation – and the pattern of tiny variations in its temperature from one direction to another (all observed with great precision by the Cosmic Background Explorer (COBE) satellite and by later ground-based and balloon-borne observations): all confirm the general picture of the hot early Universe.

This overall scenario of expansion from a hot early state to a cooler present state is usually referred to as the 'Big Bang' model and is accepted as the working picture for the evolution of the Universe by almost all cosmologists.

One of the curious features of the Universe is the way in which it presents us with an environment which is superficially extremely hostile to life. However, appearances can be deceptive. We know that the Universe is expanding and therefore its huge size is a consequence of its great age. Any universe which contains the building blocks for biological complexity must be old enough for stars to form and generate the elements on which biochemistry is based. This requires elements heavier than hydrogen and helium, which are formed during the first three minutes of the Big Bang. These heavier elements, like carbon, are made by nuclear reactions in the stars. When stars die they are dispersed into space and ultimately find their way into planets and people. This process of nuclear alchemy is long and slow. It takes billions of years to run its course. Thus a universe that contains 'observers' must be billions of years old and hence billions of light-years in size. These are necessary conditions for life to be possible. Further consequences follow. The large size of a habitable universe ensures that it has a very low average density and so galaxies and stars are widely dispersed. Outposts of life are likely to be separated by vast astronomical distances, ensuring that development occurs in isolation from other outposts of life at least until technical knowledge is very sophisticated. The large amount of expansion also ensures that the Universe is very cold. This, in turn, means that the night sky appears dark. There is too little energy density in the Universe to make it bright. Thus, universes that meet the necessary conditions for life are big and old, dark and cold.

One might speculate that these aspects of the universes (which should be universal features for observers everywhere) play a significant role in shaping our religious and philosophical impressions of the Universe and assessing our place within it. Many philosophers have appealed to the vastness and sparseness of space as evidence for a dysteleological worldview that sees the Universe as fundamentally hostile to life. Yet the discovery of the expansion of the Universe shows how subtle this matter is. Those aspects of the Universe which, to some, appear so obviously in conflict with any interpretation of the Universe as hospitable for life, turn out to be crucial features that are necessary for a universe to support complexity, life and consciousness of any known sort.

It is worth looking a little more closely at the emptiness of space. If we were to smooth out all the matter in the Universe into a smooth sea of evenly spaced particles, there would be only about one atom in every cubic metre of space – a far better vacuum than could ever be created in a laboratory on Earth. If we allow the matter to be clumped then this density would allow about one Earth-sized planet in every cube of side 10 light years, one star in every cube of side 1 000 light years, one galaxy of 10^{11} stars in every cube of side 10 million light years. Faced with these figures we begin to see why it is unsurprising that there are so few extraterrestrial lifeforms about: in a universe that is big enough and old enough to give rise to life the average distances between planets and stars are necessarily huge.

Space had small beginnings

As we look more closely at the expansion of the Universe we find that it is delicately poised, expanding very close to the critical dividing line that separates universes which expand fast enough to overcome the pull of gravity and keep going forever from those which will ultimately reverse into a state of global contraction and head toward a cataclysmic Big Crunch. Indeed, so close are we to this critical divide that our observations cannot tell us for sure what the long-range forecast is. However, it is the very proximity of the expansion to the divide that is the big mystery: a priori it seems highly unlikely to arise by chance. Again, this is not totally unexpected. Universes that expand too fast are unable to aggregate material into galaxies and stars, so the building blocks of complex life cannot be made. By contrast, universes that expand too slowly end up collapsing into contraction before the billions of years needed for stars to form have passed. Only universes that lie rather close to the critical divide

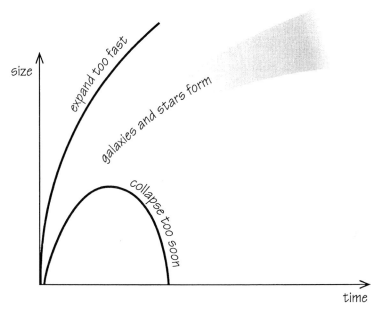

FIGURE 8.6 Open, critical and closed Universes. Universes that expand too slowly
will collapse back to a Big Crunch before galaxies can form. Universes that expand
too quickly do not allow islands of matter to condense out into galaxies. Universes
that allow galaxies and stars to form must lie close to the critical divide which
separates these two Universe types.

can live long enough and expand gently enough for the stars and planets to
form. It is no accident that we find ourselves living billions of years after the
apparent beginning of the expansion of the Universe and witnessing a state of
expansion that lies close to the critical divide (see Fig. 8.6).

Since 1980, a new cosmological theory has provided an explanation for why
the universe displays proximity to flatness and very large size. This 'inflationary'
theory of the very early universe includes a historical interlude called 'inflation'.
It creates a slight gloss on the simple picture of an expanding universe. But this
gloss has huge implications. The standard Big Bang picture of the expanding
universe, that has been with us since the 1920s, has a particular property: the
expansion is always decelerating. No matter whether the universe is destined
to expand forever, or to collapse back to a Big Crunch, the expansion is always
being decelerated by the gravitational attraction exerted by all the material in
the Universe.

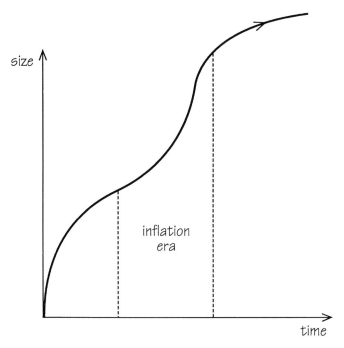

FIGURE 8.7 Inflation occurs when the expansion of the Universe accelerates for some interval of time. This must occur very early in its history, long before the appearance of stars and galaxies.

It had always been assumed that this is an attractive force. But in the 1970s, particle physicists began to find that their theories of how matter behaved at high temperatures contained a collection of matter fields, called *scalar fields*,[1] whose gravitational effect upon each other can be repulsive. If those fields were to become the largest contributors to the density of the Universe at some stage in its very early history, then the deceleration of the Universe would be replaced by a surge of acceleration. Remarkably, it appeared that if these scalar fields do exist, then they invariably come to be the most influential constituent of the Universe, and their influence only ceases when they decay away into ordinary matter and radiation.

---

[1] Scalar fields possess only a magnitude but no directional information. They can change with time or location or they can remain constant. For example, the temperature of the atmosphere is a scalar field but the wind velocity is not because at each point it has a direction as well as a magnitude.

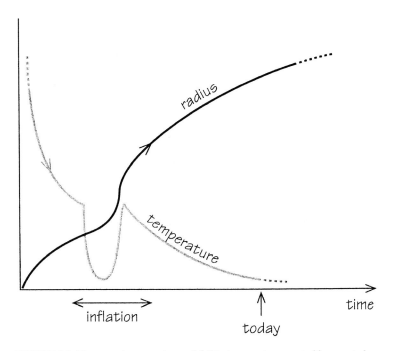

FIGURE 8.8 The surge in expansion and fall in temperature created by a period of inflation in the early Universe. When inflation ends a complicated sequence of events, involving the decay of the forms of matter driving the inflation, heats up the Universe. Subsequently, it cools steadily as it resumes expanding at a slower decelerating rate

The inflationary universe theory is simply the proposal that a brief period of accelerated expansion occurred in the very early history of the universe (Fig. 8.7).

When the inflating scalar field decays away, its energy heats up and the expansion resumes its usual decelerating expansion. The possible change in the scale of space and the temperature of radiation during the whole inflationary interlude is shown in Fig. 8.8.

This brief inflationary episode sounds innocuous but can solve many long-standing cosmological problems. It enables us to understand why our visible universe is expanding so close to the critical divide that separates open from closed universes. This is surprising because any deviation from the divide grows steadily with the passage of time. This divergent tendency is a consequence of the attractiveness of the gravitational force. If gravity is repulsive and the

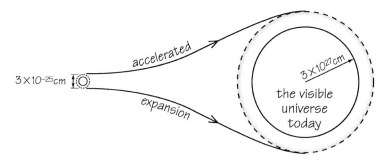

FIGURE 8.9 Inflation grows a region bigger than the visible part of the Universe today from a region small enough to be coordinated by light signals near the beginning of the expansion. This provides an explanation for the uniformity of the visible universe today, its large size, and its proximity to the critical expansion rate.

expansion accelerates, then it is driven ever closer to the critical divide. Inflation can explain why our visible universe expands so close to the critical divide.

If inflation occurred, the whole visible universe around us today will have expanded from a region that is much *smaller* than it would have originated from had the expansion always decelerated, as in the non-inflationary Big Bang theory. The smallness of our inflationary beginnings has the nice feature of offering an explanation both for the high degree of uniformity that exists in the overall expansion of the universe, and for the very small non-uniformities seen first by NASA's COBE satellite. These are the seeds that subsequently develop into galaxies and clusters. (see Fig. 8.9).

If the Universe accelerates, then the whole of our visible universe can arise from the expansion of a region that is small enough for light signals to traverse at very early times. Any internal irregularities get smoothed out very quickly. In the non-inflationary Big Bang theory the situation was very different. Our visible part of the universe had to emerge from a region 10^{25} times bigger than light rays can cross at that time. It was therefore a complete mystery why our visible universe looks so similar in every direction and from place to place on the sky to within one part in 100 000 as observations have shown. Its present uniformity means that it must have emerged from a tiny region that was very smooth. But at any early time, a region of the right size to form our visible part of the universe is always far too big for light signals to have crossed since the expansion began. It could not have smoothed out any irregularities it began with.

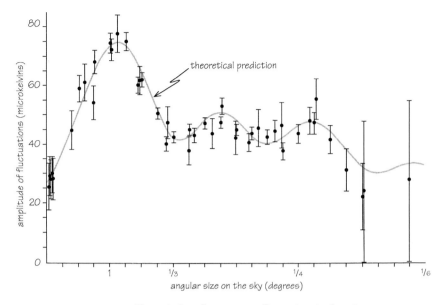

FIGURE 8.10 The variation of temperature fluctuations in the microwave background radiation found by the Boomerang project. A fit to the data by an almost critical expanding Universe's predictions for these fluctuations is shown. The angular location of the first peak in the fluctuation is our most sensitive probe of the total density of the Universe.

The tiny region which expanded to become our visible universe would not have started out perfectly smooth. That is impossible. There must have been some tiny level of random fluctuation present. Heisenberg's Uncertainty Principle requires it. Remarkably, a period of inflation stretches these essential fluctuations to very large astronomical scales, where they have been seen by the COBE satellite. During 2002, they were subjected to minute scrutiny by the Microwave Anisotropy Probe satellite (MAP) launched by NASA. If inflation occurred, the observed fluctuations should have a very particular signature. So far, the data taken by COBE are in very good agreement with the predictions, but the really decisive features of the signature appear on angular scales too small for COBE to have resolved. The new satellite observations, aided by increasingly accurate measurements of smaller portions of sky from the Earth's surface, will decide this question experimentally.

In Fig. 8.10, we see a typical prediction from an inflationary universe model for the variation of the fluctuation intensity with angular scale, together with

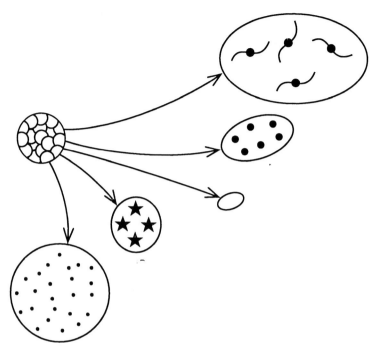

FIGURE 8.11 Chaotic inflation.

the observational data taken by Boomerang, a high-flying balloon experiment, near the Earth's surface over the Antarctic. Satellite observations will render the experimental uncertainties smaller than the thickness of the theoretical curve and provide an inescapably powerful test of particular inflationary models of the very early universe. It is remarkable that these observations are providing us with a direct experimental probe of events that occurred when the universe was only about 10^{-35} seconds old.

Inflation implies that the entire visible universe is the expanded image of a region that was small enough to allow light signals to cross it at very early times. Beyond the boundary of that little patch lie many (perhaps infinitely many) other such causally connected patches which will all undergo varying amounts of inflation to produce extended regions of our Universe that lie beyond our visible horizon today. This leads us to expect that our Universe possesses a highly complex spatial structure and the conditions that we can see within our visible horizon, about 15 billion light years away, are unlikely to be typical of those far

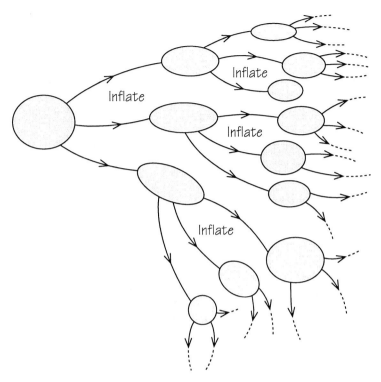

FIGURE 8.12 Eternal self-reproducing inflation.

beyond it. This complicated picture is usually termed 'chaotic inflation', and is shown schematically in Fig. 8.11. It is as if a foam of bubbles is expanded so that some bubbles become large, while others remain small. We have to find ourselves in one of the larger ones that expand for billions of years.

It has always been appreciated that the Universe might have a different structure beyond our visible horizon. However, prior to the investigation of inflationary universe models this was always regarded as an overly positivistic possibility, often suggested by pessimistic philosophers, but which has no positive evidence in its favour. The situation has changed: the chaotic inflationary universe model gives a real reason to expect that the Universe beyond our horizon differs in structure from the part that we can see.

It was then realised that the situation is probably even more complicated. If a region inflates then it necessarily creates within itself, on minute length scales, the conditions for further inflation to occur from the many sub-regions

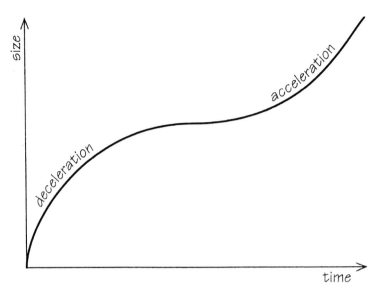

FIGURE 8.13 The repulsive gravitational effect of the cosmic vacuum on the expansion of the Universe. When it becomes larger than the attractive inverse-square force of gravity it causes the expansion of the Universe to switch from deceleration to acceleration. This changeover appears to have occurred in the last 2 billion years.

within. This process can continue into the infinite future with inflated regions producing further sub-regions which inflate, which in turn produce further sub-regions that inflate, and so on . . . *ad infinitum*. The process has no end. It has been called the 'eternal' or 'self-reproducing' inflationary universe, illustrated schematically in Fig. 8.12.

The self-reproducing character of the eternal inflationary universe seems to be an almost inevitable by-product of the sensitivity of the evolution of a universe to the inevitable presence of small quantum fluctuations in energy from place to place.

The origami of space

The phenomenon of inflation is expected to happen when the universe is very young, just about 10^{-35} seconds old. Such times are as bizarre as the huge times, like 10^{17} seconds that give the present age of the universe. Clearly, anthropomorphic units like 'seconds' were not designed for cosmological problems. There are other humanly devised measures of time which might alleviate the problem of convenience but they would still fail to give us a feeling that we had got the right superhuman temporal perspective. Fortunately, there is a measure of time that is intrinsically defined by the forces of Nature and the structure of space-time that enables us to create the right perspective. In 1870 an Irish physicist, George Johnstone Stoney (also famous for both naming and predicting the value of the charge on an electron), realised that the constants of Nature, e, G and c, could be combined to create units of mass, length and time that were independent of human standards. Thirty years later, Max Planck rediscovered this idea and expressed such 'natural units' using the physical constants h, c and G. The resulting Planck–Stoney units of length, L_{pl}, and time, T_{pl}, are the only quantities with units of length and time that can be made from these fundamental constants which govern quantum, relativistic and gravitational phenomena respectively:

$$L_{pl} = (Gh/c^3)^{1/2} = 4.1 \times 10^{-33} \text{ centimetres}$$
$$T_{pl} = (Gh/c^5)^{1/2} = 1.3 \times 10^{-43} \text{ seconds}$$

These very small scales of distance and time mark the boundaries below which our current theories of gravity and quantum reality fail. They must be replaced by a new theory of quantum gravity. In effect, at times earlier than 10^{-43} seconds the entire universe is dominated by quantum mechanical uncertainty. No one knows what the structure of space will be like if we could probe on scales smaller than L_{pl} – perhaps knotted or chaotically interconnected in some complicated way.

In these natural units the visible universe is now 10^{60} Planck-times old and 10^{60} Planck-lengths in size. This is an objective way of saying that the universe is big and old without recourse to comparisons with human measuring artifacts. One way of looking at what inflation has done for the universe is to regard the Planck-length as the natural size for a space under the influence of gravitational forces. Inflation can make the universe become much bigger, in our case more than 10^{60} times bigger.

These numbers are still very large and it may be helpful to bring them closer to the imaginable in the following way. Suppose we take a piece of A4 paper and fold it in half, then in half again. You may be surprised to find that you can't fold it in half more than 7 times. Halving moves very quickly. If you could have folded the paper in half 30 times it would be down to the size of a single atom;

fold it 47 times and you are smaller than an atomic nucleus; fold it 114 times (a lot of foldings but not unimaginable) and you are down to the Planck length where quantum gravity (and quantum origami) rules. To make the link with the scale of the whole visible universe, imagine doubling the size of the A4 paper 90 times, making it 10^{27} centimetres in size. Just 204 paper-foldings divides the smallest and the largest dimensions of space in the physical universe.

Space faces an unpredictable future

The expansion of space suggests two quite different long-range forecasts, both equally pessimistic. The closed future that returns to a Big Crunch of high temperature and density holds out little hope for the long-term survival of life or information processing unless it can base itself upon processes intrinsic to the curvature of space. There is the speculative possibility that such universes might 'bounce' back into a state of expansion and run through a sequence of oscillations, perhaps each cycle differing from its predecessors in some way. By contrast, open universes that expand forever face a future of ever-diminishing free energy resources, falling temperatures, dissolution and decay. These two scenarios are rather familiar; but in the last few years another possibility has become more likely.

Observers have used powerful ground-based telescopes to monitor nearly 100 pieces of the night sky, each containing about 1000 galaxies, at the time of the New Moon, when the sky is particularly dark. They return three weeks later and image the same fields of galaxies, looking for stars that have brightened dramatically in the meantime. They are looking for far-away supernovae: exploding stars at the ends of their life cycles. With this level of sky coverage, they will typically catch about 25 supernovae as they are brightening. Having found them, they follow up their search with detailed observations of the subsequent variation of the supernova light, watching the increase in the brightness to maximum and the ensuing fall-off back down to the level prior to explosion.

This detailed mapping of the light variation of the supernovae enables the astronomers to check that these distant supernovae have the same light signature as ones nearby that are well understood. This family resemblance enables the observers to determine the relative distances of the distant supernovae with respect to the nearby ones from their apparent peak-brightnesses, because their intrinsic brightnesses are roughly the same. This gives a powerful new method of determining the distances to the supernovae to go with the usual Doppler shift measurements of their spectra from which their speeds of recession are found. This in turn gives a new and improved version of Hubble's law of expansion out to very great distances.

The result of these observations by different international teams of astronomers is to provide evidence that the universe is accelerating (see Fig. 8.13). The striking feature of the observations is that they require the existence of the 'cosmological constant', sometimes called the cosmic vacuum energy. It behaves like the presence of another type of gravitational force that is *repulsive* rather than attractive. It was this type of repulsive force that appeared temporarily to inflate the Universe during its very early stages. The contribution of this vacuum energy to the expansion of the Universe today is about 73 per cent of its total energy density, the other 27 per cent being in the form of matter and radiation.

If the existing observations continue to be confirmed then they are telling us something very dramatic and unexpected: the expansion of the Universe is currently controlled by the cosmic vacuum stress and is *accelerating*. The implications of such a state of affairs for our understanding of the vacuum and its possible role in mediating deep connections between gravity and the other forces of Nature are very great.

But the most dramatic effect of the vacuum energy is still to come: its domination of the universe's future. The vacuum energy stays constant while every other contribution to the density of matter in the Universe – stars, planets, radiation, black holes – is diluted away by the expansion. If the vacuum energy has recently started accelerating the expansion of the Universe, its domination will grow overwhelming in the future. The Universe will continue to expand and accelerate forever. The temperature will fall faster, the stars will exhaust their reserves of nuclear fuel and implode to form dense dead relics of closely packed cold atoms or concentrated neutrons, or large black holes. Even the giant galaxies and clusters of galaxies will eventually follow suit, spiralling inwards upon themselves as the motions of their constituent stars are gradually slowed by the outward flow of gravitational waves and radiation. All their stars will be swallowed up in great central black holes, growing bigger until they have consumed all the material within reach. Ultimately, all these black holes will evaporate away by the quantum tunnelling of energy out through their horizons. They will disappear in a final explosion of particles and radiation, leaving the Universe a less interesting place than before.

The most fascinating thing about the cosmic vacuum energy is that, ultimately, it wins out over all other forms of matter and energy in the struggle to determine the shape of space and the rate of expansion of the Universe. No matter what the structure of the Universe in its earlier days, it will approach the most symmetrical possible expanding space. Everything gets smoother and smoother; all differences in the rate of expansion from one direction to another are expunged at a rapid rate; no new condensations of matter can appear out of the cosmic matter distribution; local gravitational pull has lost its last battle with the overwhelming repulsion of the cosmic vacuum force.

This has important consequences for any consideration of 'life' in the far future. If life requires information storage and processing to take place in some way then we can ask whether the Universe will always permit these things to occur. When the vacuum energy is not present, there are a range of exotic possibilities open for these rather basic lifeforms (the artificial intelligentsia?) to perpetuate themselves. This does not mean that life in any shape or form *will* survive for ever, let alone that it *must* survive forever, merely that it is logically and physically possible given the known laws of physics in the absence of a universal vacuum energy.

However, if the vacuum energy exists then everything changes – for the worse. All evolution heads inevitably for a state of uniformity characterised by the maximally symmetric accelerating universe. Information processing cannot continue forever: it must die out. There will be less and less utilisable energy available as the state of the material universe is driven closer and closer to a state of uniformity. Once the vacuum energy begins to accelerate it, the Universe looks destined for a lifeless far future. Eventually, the acceleration leads to the appearance of communication barriers. We will be unable to receive signals from sufficiently remote parts of the Universe. It is as if we are living inside a huge black hole. The part of the Universe that can affect us (or our ancestors), and to which we can send signals, will be finite.

Space may have more dimensions than three

During his early career, the great German philosopher Immanuel Kant was more interested in science than in philosophy. He was a great admirer of Newton and his laws of gravity and motion and applied himself to understanding them in different ways and to applying them to astronomical problems like the origin of the solar system. Kant's most imaginative idea, sparked by

Newton's work, was to pose the question 'why does space have three dimensions?' Kant had noticed a very profound thing: that Newton's famous inverse-square law of gravity was intimately connected with the fact that space has 3 dimensions. If space had 4 dimensions then there would be an inverse-cube law, if it had 100 dimensions there would be an inverse-99th-power law of gravity. In general, an N-dimensional world exhibits a force law for gravity which falls off as the $(N-1)$th power of distance. Consider a mass located at a point. Now surround it by a spherical surface. The lines of force following the force of attraction toward the mass point all intersect the spherical surface. It is the area of this surface that tells us what the inverse-power of distance is that the force obeys. In three-dimensional space the spherical surface is two-dimensional and has an area proportional to the square of its radius. Likewise in N-dimensional space the sphere has a surface area threaded by the force lines that is proportional to its radius to the (N-1)th power.

Kant used this observation to 'prove' that space must have 3 dimensions because of the existence of Newton's inverse-square law of gravitational force, suggesting that if God had chosen an inverse-cube rather than an inverse-square law of gravitational force then a universe of different dimensions – 4 – would have resulted. Today, we would regard this as getting the logic back to front: it is the three-dimensionality of space that dictates the inverse-square force laws in Nature. Kant went on to speculate about some of the theological and geometrical aspects of extra dimensions, and saw that it might be possible to study the properties of these hypothetical spaces by mathematical means:

> A science of all these possible kinds of space would undoubtedly be the highest enterprise which a finite understanding could undertake in the field of geometry . . . If it is possible that there could be regions with other dimensions, it is very likely that a God had somewhere brought them into being. Such higher spaces would not belong to our world, but form separate worlds.

The nature of universes with different dimensions of both space and time has been explored by a number of scientists. Just as when considering universes with other dimensions of space with one dimension of time so we assume that the laws of Nature keep the same forms but permit the numbers of dimensions of space and time to range freely over all possibilities. The situation is summarised in the picture shown in Fig. 8.14. The chequerboard of all possibilities

195

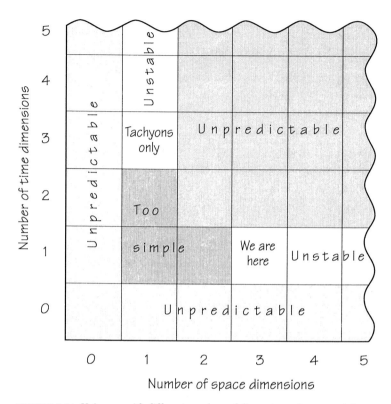

FIGURE 8.14 Universes with different numbers of dimensions of space and time have unusual properties that do not look conducive to complex information processing and life except when there are one dimension of time and three dimensions of space.

can be whittled down dramatically by the imposition of a small number of reasonable requirements that seem likely to be necessary for information processing and 'life' to exist. If we want the future to be determined by the present then we eliminate all those regions of the board marked 'unpredictable'. If we want stable atoms, planets and stars to exist then we cut out the strips marked 'unstable'. Excluding worlds in which there is only faster-than-light signalling we are left with our own world of 3 + 1 dimensions of space plus time along with very simple worlds that have 2 + 1, 1 + 1, and 1 + 2 dimensions of space plus time. Such worlds are usually thought to be too simple to contain living things. For example, in 2 + 1 worlds there are no gravitational forces between masses and there is a simplicity imposed on designs that challenges any attempt to evolve complexity. Worlds with more than one time are hard to imagine and appear to offer many more possibilities. Alas, they seem to offer so many possibilities that the elementary particles of matter are far less stable than in worlds with a single time dimension. Protons can decay easily into neutrons, positrons and neutrinos and electrons can decay into neutrons, antiprotons and neutrinos. The overall effect of extra time dimensions is to make complex structures highly unstable unless they exist at extremely low temperature.

When we look at worlds with dimensions of space and time other than $3 + 1$ we run into a striking problem. Worlds with more than one time dimension do not allow the future to be predicted from the present. In this sense they are rather like worlds with no time dimension. A complex organised system, like that needed for life, would not be able to use the information gleaned from its environment to inform its future behaviour. It would remain simple, too simple to evolve.

If the number of dimensions of space or time had been chosen at random and all numbers were possible then we would expect the number to be a very large one. It is very improbable that a small number is chosen (there are so many more larger ones). However, the constraints imposed by the need to have 'observers' to talk about the problem mean that all possibilities are not available and a three-dimensional space is forced upon us. All the alternatives will be barren of life: too simple, too unstable, and too unpredictable for complex observers to evolve and persist within them. As a result we should not be surprised to find ourselves living in three large dimensions subject to the ravages of a single time. There is no alternative space time habitat for life.

Superstring theories are the only current theories of physics which do not lead to internal contradictions or to predictions that measurable quantities have infinite values when gravity is merged with the other forces of Nature. Yet these appear to require the Universe to have many more dimensions of space than the three that we habitually experience. The original string theories required the Universe to have either nine or twenty-five dimensions of space! Since we see only three dimensions, perhaps lots of others are hiding somewhere. A process must be found which allows three (and only three) of the total number of dimensions of space to grow very large while the rest remain trapped at the

Planck scale of size, where their effects are imperceptible to us. In fact, when one looks more closely, it turns out to be rather natural that *all* the dimensions stay trapped at the Planck size. The conundrum is how three of them have become so much bigger – 10^{60} times bigger than the Planck size, in fact. What is required is a process which leads to the selective inflation of only three of the dimensions. At present no such selective process is known. If it exists, it might be random in character, so that a choice of three large dimensions was a historical accident, like the number of planets in the solar system. Alternatively, there might be a deep reason why three, and only three, dimensions can inflate.

The true constants of Nature are really framed in the total number of dimensions of space. If a complicated physical process leaves only three of them expanding then our constants of physics are just three-dimensional shadows of the true constants, which live in the full number of dimensions. If the extra dimensions were to change their size then our 'constants' of Nature would change at exactly the same rate.

One other escape from the consequences of extra dimensions and variations in constants is currently being explored in great detail. It is that only the gravitational force 'sees' and influences all the dimensions of space. The other, non-gravitational forces – electromagnetism, weak radioactive, and strong nuclear – only act in three dimensions. This would explain why gravity is so much weaker than the other forces of Nature. It also ensures that changes in the extra dimensions do not bring about variations in the constants of Nature that govern the non-gravitational forces. Time will tell whether this idea will be productive and lead to testable cosmological predictions.

And there may be more to it than meets the eye

We can easily fall into the habit of thinking that our observations characterise the entire Universe rather than just the part of it we can see. We must distinguish between two meanings of 'universe'. There is the *Universe* with a capital 'U' – that is, everything there is. This may be finite, or it may be infinite. But there is also something smaller that we call the *visible universe*. This is a spherical region centred on us, about 14 billion light years in radius, from within which light has had time to reach us since the Universe began (see Fig. 8.15). It contains all that we could possibly see of the Universe today with perfect measuring instruments of unlimited sensitivity.

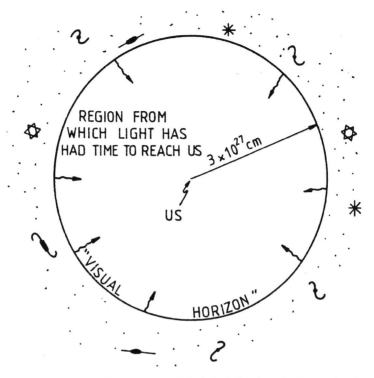

FIGURE 8.15 The visible universe is a finite spherical region of radius equal to the distance that light can travel in the time since the expansion began. The entire Universe extends far beyond this visible horizon, perhaps to infinity.

The first lesson we draw from this simple observation is that astronomy can only tell us about the structure of the visible universe. We can know nothing of what lies beyond its horizon. So, while we might be able to say whether our visible portion of space has certain properties, we can say nothing about the properties of space as a whole unless we smuggle in an assumption that the Universe beyond our horizon is similar in character to the visible universe within our horizon. This prevents us from making any testable statements about the initial structure, or the origin, of the whole Universe. The complexities of the geography of space that the inflationary universe scenarios create for us mean that for the first time we have to regard it as very probable that the visible universe has a rather different structure from that which lies over the horizon. There is more to space than meets the eye.

John D. Barrow

FURTHER READING

Barrow, J. D., *The Origin of the Universe*, London: Orion, 1994.

Barrow, J. D., *The Book of Nothing*, London: Jonathan Cape, 2000.

Barrow, J. D., *The Constants of Nature*, London: Jonathan Cape, 2002.

Barrow, J. D. and Tipler, F. J., *The Anthropic Cosmological Principle*, Oxford: Oxford University Press, 1986.

Greene, B., *The Elegant Universe*, London: Vintage, 2000.

Guth, A., *The Inflationary Universe*, Reading: Add. Wesley, 1997.

Linde, A., 'The self-reproducing inflationary universe', *Scientific American* **5**, (1994) vol. 32.

Rees, M. J., *Just Six Numbers*, London: Orion, 2000.

Smoot, G. and Davidson, K., *Wrinkles in Time*, New York: Morrow, 1994.

Tropp, E. A., Frenkel, V. and Chernin, A. D. *Alexander Friedmann: The Man who Made the Universe Expand*, Cambridge: Cambridge University Press, 1993.

Notes on contributors

Neal Ascherson was educated at King's College, Cambridge, where he read History. He has had a distinguished career as a political journalist and historian. His books include *The Polish August* (1981), *Black Sea* (1995) and *Stone Voices: The Search for Scotland* (2002). He has also been active in promoting public archaeology. Awards and honours include the Polish Order of Merit and the George Orwell Award.

John D. Barrow received his doctorate in Astrophysics from Oxford University in 1977, and held positions at Oxford and Berkeley before arriving at Sussex University in 1981. In 1999 he took up a new appointment as Research Professor of Mathematical Sciences at Cambridge and Director of the Millennium Mathematics Project, a new initiative to improve the understanding and appreciation of mathematics and its applications amongst young people and the general public. He is the author of fifteen books and a play, *Infinities*. He has lectured on cosmology at the Venice Film Festival, 10 Downing Street, Windsor Castle and the Vatican Palace.

Char Davies was originally a painter and filmmaker, and then became involved in digital technology in the mid 1980s, becoming a founding director of the 3D software company Softimage. Davies has exhibited her work at museums and galleries around the world, and has written and lectured widely. She has received numerous awards, most recently an Honorary Doctorate of Fine Arts from the University of Victoria, British Columbia. She is a Visiting Scholar at the University of California at Berkeley, and a Ph.D. Fellow in the Philosophy of Media Arts at CAiiA, the Centre for Advanced Inquiry in the Interactive Arts, University of Wales College, Wales. She is currently developing new work through her Montreal-based research company Immersence.

Karen Emmorey is a senior scientist at the Salk Institute, USA. Her research into signed languages has furnished new insights into the nature of human language, the relation between language and spatial cognition, and the determinants of brain organisation for language. Among her recent books are *Language, Cognition, and the Brain: Insights from Sign Language Research* (2002) and, as co-editor with Judy S. Reilly, *Language, Gesture, and Space* (1995).

Susan Greenfield is Fullerian Professor of Physiology at Oxford University and Director of the Royal Institution of Great Britain. She was both an undergraduate and graduate at Oxford, but has subsequently spent time in postdoctoral research at the Collège de France, Paris, and at the New York University Medical Centre, New York. In 1998 she was awarded the Michael Faraday medal by the Royal Society and in 1999 was elected to an Honorary Fellowship of the Royal College of Physicians. A prolific author and frequent television and radio presenter, she is also involved in science policy and has given a consultative seminar to the Prime Minister on the future of science in the UK. Her latest book is *Tomorrow's People: How 21st-century Technology is Changing the Way We Think and Feel* (2003).

Jeffrey Hoffman gained his doctorate in Astrophysics from Harvard University in 1971, followed by post-doctoral work at Leicester University between 1972 and 1975. He joined NASA in 1978. As an astronaut with NASA, he flew five missions aboard the Shuttle, including the 1993 mission to repair the Hubble Space Telescope. He left the astronaut programme in July 1997 to become NASA's European Representative in Paris, where he served until August 2001. After that, he was seconded by NASA to the Massachusetts Institute of Technology, where he is a Professor in the Department of Aeronautics and Astronautics. He is engaged in several research projects using the International Space Station and teaches courses on space operations and design.

Robert Howell received his first degree, in Mechanical Engineering, from the University of Sussex. He completed research for a Ph.D. in Turbine Aerodynamics at the Whittle Laboratory of Cambridge University and in 1999 he was elected a Research Fellow of Darwin College. He has acted as an aerodynamics consultant to RollsRoyce Aerospace and others and is now a Senior Aerodynamicist with Siemens Power Generation.

Lisa Jardine is Director of the AHRB Centre for Editing Lives and Letters, Professor of Renaissance Studies at Queen Mary, University of London, and an Honorary Fellow of King's College, Cambridge. She has published extensively on the Renaissance and on early modern history of science. Her most recent books include *Global Interests: Renaissance Art Between East and West* (2000, with Jerry Brotton), *On A Grander Scale: The Outstanding Career of Christopher Wren* (2002) and *The Curious Life of Robert Hooke: The Man Who Measured London* (2003).

Daniel Libeskind first studied music before receiving his professional architectural degree at the Cooper Union (New York) in 1970. He took a postgraduate degree in History and Theory of Architecture at the School of Comparative Studies at Essex University in 1972. An experienced teacher of architecture, he first came to public prominence when he won the competition to design the Jewish Museum in Berlin in 1989. It opened to the public in September 2001. Most recently, in July 2002, his Imperial War Museum North in Manchester opened. He is presently designing and constructing the Spiral Extension to the Victoria and Albert Museum, and his competition entry was selected for the World Trade Center Site in New York.

François Penz is an architect by training and the Director of Cambridge University Moving Image Studio (CUMIS) and the co-Director of the Architecture and the Moving Image M.Phil. and Ph.D. programmes. He is a fellow of Darwin College. He contributed to and co-edited the book *Cinema and Architecture* (1997), and writes regularly on the subject of cinema and architecture.

Gregory Radick is Lecturer in History and Philosophy of Science at the University of Leeds. In 1999–2000 he was Charles and Katharine Darwin Research Fellow at Darwin College, Cambridge University, where he received his doctorate in History and Philosophy of Science in 2001. He is co-editor, with Jonathan Hodge, of *The Cambridge Companion to Darwin* (2003). He is presently completing a book on the history of studies of primate communication and the evolutionary origins of language.

Index